STOREY'S GUIDE TO RAISING RABBITS

FOURTH EDITION

Storey's Guide to
RAISING
RABBITS

Breeds ▪ Care ▪ Housing

BOB BENNETT

Storey Publishing

The mission of Storey Publishing is to serve our customers by
publishing practical information that encourages
personal independence in harmony with the environment.

Edited by Rebekah Boyd-Owens, Sarah Guare, and Deborah Burns
Art direction and book design by Cynthia N. McFarland
Cover design by Kent Lew
Text production by Erin Dawson

Front cover photograph of Checker Giant/New Zealand Cross by © Jason Houston
Interior photography credits appear on page 242
Illustrations © Elayne Sears, except for those on pages 54, 62, 64, 65, 68, and 69
Rabbit papers and certificates courtesy of the author

Expert reader: Lindsay Benoit
Indexed by Christine R. Lindemer, Boston Road Communications

Storey Publishing
210 MASS MoCA Way
North Adams, MA 01247
www.storey.com

Printed in the United States by Versa Press
10   9   8   7

Library of Congress Cataloging-in-Publication Data

Bennett, Bob, 1936–
  Storey's guide to raising rabbits / Bob Bennett. — [New ed.]
     p. cm.
  Previously published as: Raising rabbits the modern way, c1975.
  Includes index.
  ISBN 978-1-60342-456-1 (pbk. : alk. paper)
  ISBN 978-1-60342-457-8 (hardcover : alk. paper)
  1. Rabbits. I. Bennett, Bob, 1936– Raising rabbits the modern way.
  II. Title. III. Title: Raising rabbits.
SF453.B462 2009
636.932'2—dc22
                                        2009014200

To my mother,

who helped me to start right,

and

to my wife,

who encouraged me to continue

Contents

Preface

IN 1948, AS A BOY IN VERMONT, I started raising rabbits right because Tony Pisanelli had my best interests at heart. He made sure I began with good rabbits! My backyard hutches produced not only lots of rabbits but also spending money and time left over to spend it, not to mention a few extra dollars to put in the bank. I won a trophy and a few ribbons and earned a Boy Scout merit badge, too. And the rabbit dinners from that era are unforgettable.

Fifteen years after Tony got me going — after college, military service, and apartment life — I returned to my boyhood pursuit of rabbit raising in the backyard, and at this writing, more than 60 years later, I remain happily hot at it. After raising the animals myself and watching others try it with varying results, I'm prompted to pass along the tried-and-true as well as the newer ways I learned from Tony and many others — ways to house, feed, breed, show, and sell rabbits. My own experience includes building five rabbitries in two states and a stretch of six decades of raising these versatile animals.

When I recall my days as a Scout, I remember that my original reason for pursuing the rabbit-raising badge was the thought that the rabbits would do most of the work, leaving me time to pursue other badges as well as part-time jobs, and yes, even schoolwork. I needed 21 badges to make Eagle, my original goal. Some were mandatory, and I had those. The rabbit-raising badge was among the optionals.

I still have the uniform sash with all the badges my mother stitched to it. The twentieth badge was Rabbit Raising. She had hoped to sew on one more and be the mother of an Eagle Scout. It was not to be. My interest in pursuing another badge waned, but I raised lots more rabbits during my high-school days. The rabbits continued to do most of the work and to produce some of the spending money that high-school boys think they require.

The Boy Scout episode led to another one years later, which actually resulted in the publication of this book. A neighborhood Scout knew I raised rabbits and asked if I would be his counselor and help him earn that badge. "Sure," I told him. "Just get the booklet that outlines the requirements and we will get to work." To my surprise, it was the same booklet I had used 20 years earlier. George Templeton wrote it when he ran the United States Rabbit Experiment Station near Los Angeles, where most of the nation's meat rabbits were produced until rising real estate value dispatched them to the Ozarks in the 1950s, and President Eisenhower revoked funding for the federal facility.

Advances in feeds, tools, equipment, and management practices occurred during subsequent years; by the 1970s, that little manual really needed an update. George Templeton had passed away, so the ex-journalist in me produced a new one for the Boy Scouts. The national publicity it

received when it appeared suggested to some that I should produce a full-length book for everyone, not just Scouts.

This is it. Sort of.

For years after its publication, the mothers of Scouts across the country sewed that badge onto 80,000 uniform sashes; it was earned by Scouts who followed the requirements and instructions in the new manual, which gave my own mother some measure of satisfaction. Since then, more than 250,000 would-be and experienced rabbit raisers have read it; it has been reprinted about 60 times in a succession of expanded and updated editions as well as some translations. This new edition has been updated, revised, and expanded once again. The information is solid, the techniques tried and true. That's not to say you can't do things differently — rabbits are resilient creatures. But the methods here have led to unparalleled success.

When this book debuted, it was called *Raising Rabbits the Modern Way*. It contained only 150 pages. Originally, I drew heavily on the wisdom and experience of Tony Pisanelli and others of his generation. Over the years since then, I have visited many rabbitries, attended shows and conventions, and spoken with many groups of raisers and suppliers to the industry. I have maintained correspondence with literally thousands of readers, many of whom have contributed their expertise, tips, and techniques. You will meet several of them in these pages. My postal and e-mail addresses are in the book, and I would be happy to hear from you, too.

Because of all the good, solid information based on the experiences of so many people, I have been writing and rewriting this book for half of my life so far. I hope you will enjoy reading it. Tony Pisanelli would be pleased. I also strongly suggest that you join the American Rabbit Breeders Association. After serving the ARBA as a board member, and founding and editing its *Domestic Rabbits* magazine, I know that the organization has a lot to offer both beginning and advanced rabbit raisers. If you want to be successful with rabbits, read this book and join the ARBA.

— Bob Bennett
Shelburne, Vermont
Spring 2009

ACKNOWLEDGMENTS

I am very grateful to the following persons for their expertise and encouragement: Anthony Pisanelli; Robert Noble; Charles Lyons; Charles Maurer; Frank Miglis; Robert Densmore; Ted, Flo Ann, and Tammy Gordon; James Blyth; John Dack; Frank DelMastro; Pat Schmidt; H. John Nelson; Oren Reynolds; Robert Dubbell; Edward Peifer; Edward Stahl; Bill Dorn; Paul Posel; J. Calvin Downing, DVM; Jerry Belanger; Sam Mines; H. Joseph Hull; Ron Epstein; and Lindsay Benoit.

Sincere thanks go to each individual and organization for information used here. Particularly helpful were Pel-Freez, Safeguard Products, Inc., and Bass Equipment Company.

Walter Hard, Constance Oxley, Deborah Burns, Marie Salter, and Rebekah Boyd-Owens, my editors, made helpful suggestions that improved the manuscript. Sarah Guare and Mars Vilaubi, with their production and photo-editing expertise, contributed a great deal to the finished product, as did illustrator Elayne Sears.

I would also like to thank the many persons mentioned throughout the book, the hundreds of members of the American Rabbit Breeders Association, the New Jersey Rabbit Breeders Association, and the Green Mountain Rabbit Breeders Association, who over the years have contributed much that found its way into this book.

Finally, I'd like to extend special thanks to Glen Carr of the American Rabbit Breeders Association and to Pamela Bernardini for supplying some of the photographs for the breed gallery. Many thanks also go to Gary Bass of Bass Equipment Company and Willis Kurtz of Safeguard Products, Inc. for photographs that were the basis for several illustrations of equipment in the book.

PART I

Getting Started

1

Why Start?

TAKE ONE LOOK AT THE DOMESTIC RABBIT, and you will understand its appeal. Recognize it as an animal that thrives almost anywhere, is inexpensive to procure, and is easy to maintain. Add its legendary prolificacy, unmatched cleanliness, and steadfast refusal to bark at the moon at midnight or crow at the neighbors at 6:00 a.m. Consider, above all, its versatility, and you will readily comprehend the rabbit's perennial popularity.

When I think of versatility, I'm reminded of a fanciful creature that was born in a comic strip of the late 1940s. It was then that *Li'l Abner's* creator, Al Capp, introduced the Shmoo, a prolific and lovable little animal that was perfectly willing to be anything anyone in Dogpatch desired. If Daisy Mae wanted pork chops or steak, ice cream or cake, the Shmoo, which resembled an obese bowling pin, immediately multiplied. It became a two-legged movable feast like manna from heaven. The Shmoo always obliged. So warmed by this accommodating fantasy was faddish America that toy Shmoos soon appeared for purchase in novelty stores across the land.

Though the Shmoo was only the figment of a lively imagination, it could easily have been inspired by the rabbit, a prolific and lovable little creature that actually is many of the things the shmoo purported to be.

First of All, Good Eating

Above all, the domestic rabbit produced in backyards and in commercial rabbitries is mighty good to eat. It can be prepared and served in so many tempting ways that even Daisy Mae would favor it over the Shmoo. The accommodating rabbit also serves daily in medical research and as mittens, muffs, angora sweaters, toys, novelties, hats, collars, cuffs, and coats. Gardeners beg for its manure, which boasts unique properties that benefit trees, shrubs, lawns, flowers, and vegetables alike. Some gardeners raise them for no other reason.

The all-white, fine-grained meat of the domestic rabbit finds its way to fancy food stores, specialty mail-order companies, and supermarkets as packaged, cut-up, 2½-pound (1.1 kg) fryer-broilers, at a price competitive with that of prime and choice beef. Rabbit meat exceeds beef, pork, lamb, and chicken in protein content. It has a lower percentage of fat, less cholesterol, and fewer calories than any other meat you can buy. A small portion of rabbit goes a long way. Delicious hot or cold, fancy or plain, it can be breaded and fried, broiled, baked, or barbecued. Large rabbits often are stewed or roasted. While some compare its taste to that of chicken, you will fill up faster on an equal portion of rabbit, which has finer bones and a chewier texture.

Meat Rabbits in the United States

In the United States the area of northern Arkansas and southern Missouri contains 1,500 raisers of meat rabbits. They range from backyard producers to full-time commercial operators who serve the nation's largest processing plant, operated by Pel-Freez, Inc., in Rogers, Arkansas.

While this area, generally known as the Ozarks, boasts the greatest concentration of rabbit producers in the country, many thousands more are scattered throughout the United States. An independent research company has estimated total United States annual production of live-weight rabbit meat at 34.2 million pounds (15.4 million kg). Considering that the rabbit normally is marketed at a weight of 4 pounds (1.8 kg), we see that nearly 8.5 million rabbits are produced in the United States each year for domestic consumption.

A Favorite in Europe

Production and consumption of rabbits in the United States neverthe-less lag far behind that of Europe, where Italy, the Ukraine, France, and Spain are among the world's biggest producers and most willing consum-ers. German and British production and total consumption figures equal those of the United States, but the Britons and Germans have much higher per capita consumption rates. The Russians and Chinese are also big producers and consumers.

Wherever land is at a premium, the rabbit shines. One doe, in one hutch, can produce 70 to 95 pounds (31.8 to 43.1 kg) of edible meat each year — about nine times her own live weight. In the United States, where plenty of land is still available to plant corn and soybeans for hogs and to graze cattle, rabbit raising has not gained the attention it has elsewhere. As land for livestock becomes increasingly scarce and demand for grain for human consumption increases, rabbits may become more popular in the United States.

Rabbits Serve Science

Another significant use of rabbits is in the scientific sector. More than 2 million are used in U.S. medical and pharmaceutical laboratories each year. Many of these rabbits are raised in the Delaware-Maryland-Virginia (Delmarva) area and in Pennsylvania, New York, New Jersey, and Massa-chusetts, as well as California. But others are trucked up to the Northeast from the Ozarks to meet the demand. In fact, as biological products such as eyes, kidneys, and testicles derived from rabbits have become more and more valuable to research laboratories, meat has turned into a by-product for the biggest processors.

The Fancier's Role

Adding the meat and laboratory production figures together, you might surmise that there is a total U.S. rabbit population of 10.5 million. But that figure would exclude those in the hutches of thousands of fanciers who raise rabbits for their own consumption and for exhibition and sale as pets.

Basic to the acceptance of meat and laboratory rabbits is the availability of quality, well-bred stock. That is provided by the fanciers, who carefully guard the purity of the pedigrees of 50 recognized breeds, which range from 20-pound (9.1 kg) giants to the 2-pound (0.9 kg) dwarfs famous for popping out of magicians' hats. These fanciers and some commercial growers hold membership in the American Rabbit Breeders Association, which maintains a national headquarters in Bloomington, Illinois.

In addition to supplying choice breeding stock required by profit-conscious commercial growers, they supply pets, meat, and laboratory specimens and compete in hundreds of fairs and special exhibitions, including a weeklong national convention and gigantic show that moves to a different location each year. A recent show included 22,000 rabbits, and that's not the biggest one ever. They belong to hundreds of state and county breeder associations, as well as to national breed specialty clubs dedicated to their favorite kind of rabbit. They read scores of national, regional, and local trade and association publications, most of which are produced by volunteers from their own ranks. They also read *Domestic Rabbits* magazine, the official bimonthly publication of the ARBA.

Would You Believe 12 Million Rabbits?

ARBA membership rolls include about 30,000 adults and youths. There are doubtless an additional 200,000 unidentified backyard rabbit raisers with no affiliation whatsoever. In one county in New Jersey, for example, for every rabbit raiser who belongs to the ARBA, you will find at least a dozen small raisers who do not. One feed store's manager reports selling 100 pounds (45.4 kg) of feed a week to 50 raisers who don't belong to any rabbit clubs. I'd estimate, therefore, that about 12 million domestic rabbits exist in the United States. To find them, you have to peek into backyards, sheds, and garages in cities, small towns, and suburbs. To quantify them, all you have to do is ask the milling companies how much rabbit feed they sell. That's what I did. You could also ask the American Pet Products Manufacturers Association, which has estimated that there are more than 5.28 million pet rabbits in 2.2 million households in the United States.

Here's Where You Come In

After this brief review of the scope of the rabbit industry, it is perfectly appropriate for you to inquire where you fit into the picture. For the answer, ask yourself why, in fact, you should raise rabbits at all.

One reason might be for the simple joy of the miracle of living things, their companionship, and the pleasure of meeting their needs. You might raise them like so many pretty flowers and simply enjoy watching them bloom. Many successful rabbit raisers need no further reason for keeping their hutches filled with beautiful rabbits.

You might be more practical (if you're like me) and demand utility as well as beauty. So you might propagate rabbits to have a pair of warm, furry mittens, a delectable dish on the table, and a gleaming trophy in the showroom — even college tuition for your children, a roof over your head, or money in the bank.

Good reasons abound, but at least one is bad: to get rich quick. With hard work, good advice, and a sound business plan, many people have turned rabbits into a good living, but like many other farming activities, the rewards often are more emotional than financial.

Progress in Raising Rabbits

This book owes its existence more to progress in rabbit raising than to anything else. Some other books on the subject were written before recent, more efficient methods of housing, feeding, and managing rabbits were devised. Authors often advocated home construction of hutches from shipping crates, nail kegs, and lumber. Amazingly, such advice continues to be printed, years after it should have ceased! You will find no such recommendations on these pages.

Today, rabbit raisers have at their disposal a broad line of equipment specially designed for the animals. Complete, pelleted feed containing every nutritional element required by the rabbit is readily available nationwide. No longer must producers risk formulating their own feeds. No longer must they fool with mixed grains or possibly moldy greens in an effort to find a feed that will nourish their stock without the risk of making them sick and killing them.

You who start raising rabbits now have everything in your favor — feed, equipment, preventive medicine, and especially *sound advice*! Others have cleared the way for you, but the most important step remains, and only *you* can take it. Put your foot down! Say that this thing is worth doing well. Make that kind of commitment, and I guarantee that you will follow in the footsteps of those who are raising rabbits the very best way possible.

2

The Right Rabbit

THE BEST WAY TO START RIGHT WITH RABBITS is to start with the right rabbit. That is not as simple as it sounds. The American Rabbit Breeders Association (ARBA) recognizes about 50 individual breeds of domestic rabbits, (with new breeds or varieties recognized every couple of years) and some breeds include more than 15 color or pattern varieties. We've included photos of some of these in our Breed Gallery on page 208. The rabbit that is available in so many sizes, shapes and colors is not a wild animal but a domesticated farm species. It is not a rodent, either, but a *lagomorph*, a gnawing mammal of the order Lagomorpha. Those popular misconceptions keep surfacing and creating resistance to the rabbit. The truth is that people will not be eating a stringy, tough, wild-game animal that must be chewed cautiously to avoid swallowing birdshot.

A 40-Million-Year-Old Species?

The rabbit was domesticated several centuries ago. It might have happened first in Africa, Spain, France, Belgium, or Rome, depending on which report you read. I have come across such varying accounts that without doing some formal research, I don't know which, if any, to believe. It is said that wild rabbits were used for food in Asia at least 3,000 years ago. Geologists claim to have proven that rabbits and opos-

---

RABBIT LEGEND AND LORE

The Easter bunny legend originated with the Teutons, an ancient Germanic tribe. According to one account, a goddess changed a bird into a rabbit, and the rabbit was understandably so appreciative of this metamorphosis that when the goddess scheduled her spring festival, the lagomorph laid colored eggs for the occasion. Imagine a rabbit that could lay eggs. It boggles the mind!

The use of rabbit feet as good-luck charms dates to the following old superstition: the left hind foot of a rabbit taken into a churchyard at midnight when the moon is full will shield its owner from evil. I for one have never believed that: I mean, how much did it do for its original owner? I confess, however, to remaining absent from churchyards on clear, moonlit evenings.

---

sums are the oldest-known living mammals and date back 30 or 40 million years. Ancient Spanish caves contain paintings of rabbits. A sphinx built in Turkey in 1500 BC has been standing on the figures of two rabbits for more than 3,500 years. At any rate, we know that the rabbit, both wild and domestic, has been around for a long, long time.

American Domestics: Probably Not Belgian Hares

Domestic rabbits doubtless were raised on North American soil before the Belgian-hare boom at the beginning of the twentieth century, but that's when rabbit keeping really took off in the United States. Some people still refer to all domestic rabbits as Belgian hares (it's not actually a *hare*, a closely related wild species, but a *rabbit*), but the Belgian actually is a somewhat rare breed in the United States today.

Possibly the most important development in the history of domestic rabbits in this country occurred in 1913, when the New Zealand Red arrived on these shores. Reds were brought to the Pacific Coast by sailors who may have obtained them in New Zealand — hence the breed's name. New Zealand Reds carried better meat qualities than the Belgians and became the foundation stock for some of the most significant breeds.

A lot more history of rabbits can be found if you are so inclined, but the task at hand is to build your future with rabbits. So it is probably sufficient to say that the many breeds and varieties we have today were generally carefully, although sometimes accidentally, developed for specific purposes by fanciers.

Choose Your Market

The single most important step you will take toward starting right is to ask yourself this question: What am I going to do with my rabbits? Unless you ask and thoughtfully answer this question, it is highly likely you will become a rabbit *keeper* in every sense of the word.

Pick your purpose or your market *before* you pick your breed. Meat, laboratory, and breeding stock are the major uses of rabbits. Although ordinarily not nearly as profitable or practical, the pet market has nevertheless captured the time and energy of many raisers. And rabbit raisers also contribute Angora wool and rabbit pelts to the marketplace for making crafts and clothing.

You might suspect that I will recommend you raise only one breed for only one purpose, and you are almost right. Most breeds, however, can and will serve more than one purpose.

The Meat Market

By far, more rabbits are raised year-round for meat than for any other use. You may sell your rabbits live to a processor or slaughter them yourself for direct sale to the consumer. Live sales are the simplest because commercial, wholesale, or retail butchering is a complex operation governed by laws that require standards of sanitation, refrigeration, and ventilation that you may be unable to meet. Home butchering is simple, however, and you can sell to family, friends, neighbors, and coworkers, as long as they are willing to buy.

Before you choose a breed to sell live to a processor, find out what the processor requires. Processors often prefer a certain breed and will recommend one to you. Because the processor will be buying the young rabbits you produce, you will do well to take their advice. Processors often have weight requirements, so you'll want to produce a breed in that weight range.

I can't stress the marketing concept approach too much: *Find out what a buyer wants and deliver that product.* Some want little white rabbits, and some want big ones of any color. Ask them what they need and produce whatever that is. But don't put all your rabbits in one basket. Some processors want more rabbits in the summer, for example, while others want more in the winter — and you want to sell year-round. Deal with more than one processor if you possibly can.

New Zealand Whites and Californian

Americans raise more New Zealand Whites for meat than any other breed. Next in line are Californians, followed, not necessarily in this order, by Satins, Champagne D'Argents, American Chinchillas, and Palominos. Actually, other breeds taste just as good, but certain other considerations set some breeds apart from the rest.

The most obvious advantage of raising these breeds is that New Zealands and Californians wear white pelts. Many processors sell the pelts and prefer white because it can be dyed any color. Most rabbit fur produced in the United States is used either white or dyed, sometimes to imitate a more valuable fur.

What is more important, however, is that New Zealand Whites and Californians both have been bred to a higher degree of consistent excellence than any other breed. To the breeder, that means counting on superior feed conversion, disease resistance, fertility, litter size, and better overall performance than other breeds deliver. To processors, New Zealands and Californians mean not just white pelts but also fine bones and good *dressout* — in other words, more meat and less waste.

Florida White

In recent years the Florida White has also achieved prominence as a meat rabbit. Even though it is only half the size of the New Zealand, it has rapid early-growth characteristics and is unsurpassed pound for pound in dressout quality. It is an especially good breed for the home meat producer. Some raisers mate a New Zealand or Californian buck with a Florida White doe, with an eye toward getting somewhat larger offspring from a comparatively small doe in a compact cage. For a larger carcass

New York PR Guy and Sales Whiz, Still Raising Rabbits

J IM HAMMOND AND I once occupied offices in New York City at Second Avenue and 42nd Street, on the 20th floor of Pfizer World Headquarters, as public relations managers, helping the pharmaceutical giant polish its image and reputation. That was years ago. Now Jim is retired and raising rabbits and pygmy goats in upstate New York.

"I learned the hard way," he told me recently, "how important it is to start with good, young stock. I was in a hurry to begin years ago, so I bought some mature stock. I later found out that the local breeders don't sell anything but culls by way of mature stock.

"After a year of frustration with does who would not take care of their litters, with does almost impossible to get bred, with does prone to illness and bad dispositions, I learned that, in the long run, it's more efficient to buy young stock that a breeder isn't delighted to get rid of. After a year of trying, I didn't have one litter to show for all my efforts. The saddest part of the story is, of course, I paid a premium price for breeders, thinking they'd help get me started sooner.

"The lesson is simple. Don't go into business with someone else's castoffs. Buy young stock from a reliable breeder.

"The next lesson is to limit yourself to one or two breeds. Recently, I visited a neighbor, a teenager, who talked her dad into buying out a rabbit breeder not far from here. She asked me to come and help her organize the new additions to her rabbitry — 60 new rabbits! Of the 60, there were 8 different breeds. No wonder the guy sold out. How does one manage 8 different breeds efficiently in a single rabbitry?

"I suggested that she sell off all but two or three breeds because of the impossibility of trying to deal with so many different rabbits. Just imagine the problems of having to maintain so many bucks. One or two breeds maximum should be the rule.

"Little by little, I have managed to acquire some decent stock. There's just as much work involved in keeping inferior animals as there is in good livestock. Now I'm having a lot more fun — which for me, after all, is most of what raising rabbits is all about."

some even use a Flemish Giant buck with a New Zealand or Californian doe. Both crossbreeding techniques, as well as some other combinations, can produce excellent results.

Meat on Your Own Table

Suppose your intent is merely to raise rabbits for your own table. You still can't go wrong with New Zealand Whites or Californians. But you have many additional breeds to choose from. You don't have to meet rigid production schedules or worry that the processor will pay you less for rabbits wearing colored pelts. In fact, you might choose a rabbit of a particular color or pattern and save the pelts for craft projects of your own choosing, or even a fur coat. You don't even have to raise a particular size of rabbit.

Don't Overlook the Small Breeds

For many years I have raised smaller breeds and I can tell you that they taste just as good as the larger ones. As a matter of fact, I'm convinced that the meat of the smaller breeds can be produced more economically per pound, if you don't count the labor involved. The problem lies in finding a productive strain of small rabbits. Also true is that few processors want a smaller rabbit because of the extra time it takes to process the same amount of meat from a larger rabbit.

My favorite breed is the Tan, which weighs 4 to 6 pounds (1.8 to 2.7 kg) at maturity, about the same as the Dutch, a breed I used to raise, and the Florida White, which I have raised for a long time. These breeds don't produce New Zealand-size litters, but they eat half what the big whites do and need less space. Smaller breeds have smaller litters, ordinarily, than

The small Tans require little feed or space.

larger breeds, but smaller breeds also have finer bones and a faster early growth, producing a more finished carcass at an earlier age.

How Small Breeds Dress Out

While Tans never reach 4 pounds (1.8 kg) in eight weeks, they will weigh 3 pounds (1.4 kg) by that time. I often dress them out at ten to twelve weeks, when they deliver about as much meat as an eight-week New Zealand because there is less bone. Florida Whites perform as well or better, but to my eye, Tans are both more attractive and a greater challenge on the show table.

I don't suggest you raise Dwarfs or Polish for meat at all, because you would get little more than a pound (0.5 kg) from a mature animal. I'd remind you, though, that the woods are full of hunters who seek an even smaller animal. They fill their pots and make Brunswick stew from several squirrels.

The main thing to remember when it comes to raising meat rabbits is that when you plan to sell them live, you must meet the buyer's demands. If you plan to eat your rabbits yourself, or can develop your own customers for fresh or frozen meat, you may select almost any breed you like. For the small family, that might mean a small breed.

For the small backyard raiser, I'm sold on the concept of the small breeds. If you can keep only a few rabbits at a time, it seems to me the smaller breeds make a lot of sense. The smaller breeds have the best reputations as pets, too.

Of Pelts and Spun Wool

With the exceptions of the Angora and Rex, rabbits are not raised primarily for their coats. Fur is a by-product of the meat rabbit, and because Americans prefer a young fryer-broiler, the fur is immature and unsuitable for durable garments. At various times there has been a market for fryer fur as felt for the hat trade, when felt hats were in fashion. This market is unpredictable at best. The rabbit-fur coats worn by women and some men generally are made from the fur of European rabbits because of the preference in Europe for roaster rabbits that wear a mature, prime coat. So Europeans have cornered the commercial garment market. If you see a coat advertised as made of "French rabbit," it doesn't mean that France has better rabbits than anybody else. It means simply that the French will butcher a 10-pound (4.5 kg) rabbit for the table in midwinter when the hide is mature, thick, and prime. Remember, however, that even "French rabbit" is no match for the mink and several other furbearers with considerably thicker pelts that are made into much longerwearing garments.

In the case of angora wool, demand remains specialized but steady, and the return can be well worth the effort. Angora wool production is labor intensive. Because China has an abundance of cheap labor, the Chinese produce most of the world's Angora wool supply. Much also comes from Germany, where production techniques and mechanization

LOTS AND LOTS OF RABBITS

To make a decent profit on live meat-rabbit sales to a processor requires a high sales volume. You need to raise and sell lots of rabbits. If you go into rabbit raising for full-time income, you will need to sell at least several hundred fryers a week because you are likely to earn only a dollar or so of profit per rabbit.

Live meat rabbits consistently, and over many years, have brought only a dollar or two each over feed costs in my area. A weaned fryer, eight weeks old, requires about 16 pounds (7.3 kg) of feed in a reasonable feed/meat conversion ratio of 4:1 — that is, 4 pounds (1.8 kg) of feed to 1 pound (0.5 kg) of live rabbit. That amount of feed includes the doe but not the buck. Remember that it takes three months, from conception to market, to get that dollar or two, and you can see that it will take a lot of these fryers to make any money for you. If you hold the weanlings until they reach 5 pounds (2.3 kg), your feed costs are greater, of course.

Mind you, we are talking only about feed. We haven't yet considered the cost of stock, equipment, and any other investments you might have to make in your rabbit enterprise. True, these costs are in the category of long-term investment. But how long do you want to take to recover this cash?

of processing have reached a high degree of excellence. If you are a hand-spinner and knitter and have the time for grooming and shearing, you might well create a customer base and profit with Angora wool. At rabbit shows I often see teenage girls and young women lovingly combing and brushing their Angoras.

At times the pelts of the Rex rabbit fetch prices that make their production and sale worthwhile, but that usually requires a demand for large "roaster" rabbits, because only the mature, prime pelts of full-grown specimens will sell. The price received for the pelt alone may not cover production costs so the sale of meat is usually part of the equation. However, you might succeed with both wool and pelts, if you are willing to make a special effort.

The Laboratory Market

The laboratory market can come into play along with the meat market: some meat producers and processors sell laboratory rabbits as well. New Zealands, in particular, are preferred by laboratories. Researchers like New Zealands primarily because these rabbits can always be obtained in sufficient consistent quantity to make research results more accurate. But the Florida White also deserves more consideration as a laboratory rabbit than it has received.

Let's suppose you are considering the large-scale production of meat rabbits for live sale to a processor. I would recommend New Zealand Whites or Californians if those are what the processor buys. If you are going to sell laboratory stock as well, stick to New Zealand Whites, unless you have a commitment from a lab for another breed. Again, don't choose a breed for personal reasons — one that you think is cute, for example. Choose a breed that someone who represents a market wants to buy. Then produce the size, age, and quantity the buyer demands.

Lab Return May Be Better

Laboratory rabbits generally sell for considerably more than meat rabbits, but this market can be difficult. Not all of your production, even under the best market conditions, can be sold to laboratories. Often the order is for all one sex, perhaps females. Also, many laboratories want rabbits older than 4-pound (1.8 kg) fryers. After 4 pounds, the feed conversion ratio deteriorates. So the return on a combined meat- and laboratory-animal operation may fall only about halfway between laboratory and meat payment, minus the feed cost. You will still need a lot of rabbits to make this kind of rabbitry pay.

If you can have a big rabbitry, are a good manager, and enjoy a steady demand, you can make money. It will take you quite a long time, how-ever, because no matter what kind of rabbitry you have, you must start fairly small to start right. You can give yourself a bonus by making your own direct laboratory contract. You'll get a higher return than going through a broker. But you may have to wait to do this until you are able to be a steady enough supplier to handle it.

If managing a rabbitry to supply laboratories appeals to you, I strongly urge you to begin with New Zealand Whites or Californians — perhaps both (again, if those are what the buyer wants). If you choose one or both of these breeds and favorable market conditions prevail, you will be on your way to a successful commercial enterprise.

What about Pets?

The pet market has become huge. Affluent, single, young people with no children make up a large segment of the population and in recent years have simply gone pet crazy. The entire pet market — dogs and cats, as well as ferrets, snakes, fish, birds, and who knows what else — recently commanded $40 billion in a single year. Rabbits are a part of that.

Although unreliable and seasonal in some places, the pet-rabbit mar-ket has expanded beyond the Easter holiday. Rabbits have become Val-entine's Day and birthday gifts and year-round companions for people who might prefer dogs or cats but cannot maintain them, sometimes because of apartment or condominium rules. Selling pets has several disadvantages, however. Easter, still the time of biggest demand, requires the birth of bunnies in cold winter weather in many parts of the United States. This is the one time of year that a litter is most at risk. Pet stores and dealers demand large numbers of just-weaned bunnies on a specific date and at a low wholesale price because they have to mark them up to make a profit. At other times of the year, those pet stores willing to carry rabbits may pay you even less.

But that's only part of the problem. The rabbits often decline quickly, sometimes dying at the hands of young children who don't know when to stop squeezing, or don't understand that baby bunnies can't eat lettuce or cabbage without contracting deadly diarrhea.

The pet market, like just about everything else, has its ups and downs. When times are good, rabbits appear to many people as ideal pets. When the economy is bad, they may become the ideal meal. It should be said, however, that the sale of pet rabbits hurts their image as food, which is frustrating to many of us who raise them for that purpose. Rabbit-meat consumption rates often remain low where pet sales are high.

Here's a recent incident that serves as an example: Not long ago a well-heeled young matron with a six-year-old in tow drove to my place in a new BMW with the goal of obtaining a pet for the boy. He liked the Tans, but she ran a hand over the loin of a Florida White and remarked how solid and meaty it was, and soon the discussion turned from pets to recipes. The state of the economy dictates the thinking of a lot of people when it comes to rabbits. Back in the 1970s, when food and energy prices rose sharply, many in the United States thought of rabbits as a delectable dish. Later on, with economic prosperity, rabbits as pets gained a foothold. When economic conditions change, rabbits change with them. I told you they were versatile. Now you know they are also *adaptable*.

The Breeding-Stock Market

Profit on a meat or laboratory operation, we have seen, comes only with large volume. A small raiser, then, in my opinion, should concentrate on sales of breeding stock to prospective meat- and laboratory-stock producers and fanciers and sell only occasionally to the meat and laboratory markets — and then only if there are enough rabbits left after consigning the culls (that is, rabbits you don't want) to your own freezer.

I'm a backyard raiser with only 36 hutches in a small barn that I built. I really have no interest in anything larger. After years of raising other breeds, I chose the small breed of Tan for a number of reasons that make

> ## A PROMISING MARKET
>
> Breeding stock is a market unto itself, but it requires customers who want to raise rabbits, and there are fewer of them than there are customers who want to eat rabbits. Nevertheless, breeding stock is a promising market because there *are* people like you who want to raise rabbits. I mentioned earlier that membership in the ARBA numbers about 30,000 nationally. These members and many other thousands of people are potential customers for breeding stock. Breeding-stock demand continues to be brisk, and the outlook for continued growth is great because more and more Americans are recognizing the rewards of raising rabbits.
>
> If you raise really fine rabbits and are willing to make the effort required, you will always be able to sell breeding stock. If you start with good stock yourself, you will be able to sell breeding stock almost immediately. And what you sell will be just as good as what you bought — perhaps better.

sense to me. First, the Tan is a very fancy breed that challenges the raiser when it comes to producing top specimens. Body markings and color are important considerations. Therefore, sales of these rabbits to other raisers bring a high unit price. Also, because these rabbits consume half of what the larger breeds do, and because their cages are only about half the size of those for larger stock, it costs me very little to raise them. I can house and feed twice as many Tans or Florida Whites in the same space and with the same feed I'd need for larger rabbits. I can even raise them faster, because they mature earlier and reproduce sooner than larger rabbits.

There is a hitch, of course. I do have to advertise, and because Tans primarily are show rabbits, I have to exhibit them regularly. They must be competitive to make them marketable. Other show people pay attention and want to add winners to their herds. Showing costs little but takes time — usually on Saturdays or Sundays in the spring and fall months. For me it is an enjoyable activity, mostly because I like the company of other rabbit breeders, but also because of the fun of the competition and the tips to be gained from other breeders and the rabbit judges. Shows also make a fine family outing, and there's no denying the satisfaction of winning.

The Enticing Economics of a Small Breeding-Stock Rabbitry

Consider the economics of a small rabbitry where individual animals sell as breeding stock for a lot of money, instead of for a little, as do animals raised for meat. A breeding-stock business is advantageous to you even when you include the costs of advertising (not too high, really, and some is free!) and the sales effort that must be made. The potential per unit of space simply is much greater for the small producer of breeding stock — and, I think, even greater yet for the small producer of small breeding stock. Because not all the production will be up to breeding-stock standards, you will still have rabbits for the home freezer or even for the processor or your own customers, as well as pelts for sale or for your own use.

After having run this kind of rabbitry for many years, I'm now at the point where half of my 36 hutches contain breeding does. The rest are for bucks and young growing stock. This rabbitry has produced about 350 youngsters per year (which is nothing special — much more could be done in the same amount of space). About 275 of those have been sold for breeding stock, and the rest for meat one way or another. Few have been sold for pets, although I do give some away.

Most of the breeding stock is sold as the result of magazine advertising. I have shipped some to almost every state in the Union and to some locations beyond our boundaries. Some are sold at shows I attend annually and a few are sold locally. I like mail-order sales as a way of doing business because I can do it at times that suit my schedule.

In any case, it can be seen that the return on my small rabbitry is much greater than it would be if I tried to produce meat and laboratory stock. I should add that because I have a small rabbitry, I have more time for my many other interests; rabbits are only part of the picture.

If you are to go into the breeding-stock market, initially you will purchase a pair or more of rabbits from a breeder who sells breeding stock, as you would regardless of your intended market. The breeding-stock market is wide open when it comes to breed choices. I have already stated a case for the small, fancy breeds, but potential raisers seeking stock may be interested in one or all of the three major markets, which means that just about any breed will serve you well, from the most popular to the most obscure.

There are two ways to look at the breeding-stock market: you may select either a breed that is popular and in great demand, such as the New Zealand White, or one that is largely unavailable except from you — perhaps the Harlequin. In fact, a combination of these two may serve you well. I raised Dutch, a perennially popular small breed, for several years, while also raising Tans, then a little-known variety, to give two strings to my bow. While I worked to promote Tans, the Dutch supported the effort. Once the Tans began to gain acceptance, I was able to focus on them exclusively.

While the choice of a breed is pretty much up to you when it comes to breeding stock, one consideration is paramount. You must have purebred rabbits. While purebred stock is not always a necessity for the meat or laboratory markets, starting right, to my mind, dictates beginning with only the best available. And these are, almost without exception, purebred rabbits.

My Favorite Breeds

To help you decide, see the list below of my own favorite breeds, in descending order, based on experience that goes back to when the twentieth century was just middle-aged.

1. **The breed that people want to buy.** It could be virtually any breed, but most likely it's my next favorite — the New Zealand.

2. **The New Zealand White.** Why? Because more people buy more of them than they do any other breed. Rabbits are mostly for eating, and the New Zealand White gets eaten most often.

3. **The Tan.** Big surprise! Since I have raised Tans for 40 years, I could bore you for hours about Tans. Suffice it to say that I like their looks — and their looks are more challenging on the show table than those

of any other breed. It takes longer to judge Tans than other breeds because there are so many different characteristics to examine.

4. **The Florida White.** Though half the size of New Zealands, these rabbits dress out better than any other breed on a percentage basis. Pound for pound, they deliver more meat than any other breed and so are a great choice for the backyard raiser and for home consumption. They are small eaters in small cages, and because of their size there will be no leftovers in the fridge.

5. **The Florida–New Zealand cross.** Use a small New Zealand White buck and a large Florida White doe, and you will produce a lot of meat in a small hutch. Yes, this is a crossbreed, but to make it work right, you have to use excellent purebred New Zealands and Floridas. These are for home consumption or customers who want them.

6. **For showing, a breed that is very popular in your area.** You'll have lots of competition as well as lots of fun. Go to a few shows to determine what breed is most popular in your area.

7. **The Dutch.** This one makes a great pet rabbit and is just the right size for showing, if you want to drive yourself crazy. The markings are what make you crazy, because they are largely a matter of luck. But these are striking rabbits to have around, and the dressout on those mismarked ones is excellent.

8. **The Champagne D'Argent.** This one is an excellent meat rabbit that changes from black to what is called "French silver" as it matures.

9. **The New Zealand Red.** This one was my first breed, and I list it for sentimental reasons.

10. **The Mini Rex.** This little rabbit gives you the Rex fur without taking up a lot of space and feed. It has become one of the most popular for keeping as a pet.

THREE MARKETS MEAN SUCCESS

Sales of meat and laboratory stock, supplemented by breeding-stock sales, are the best ways I know to succeed with a larger rabbitry. It takes considerable time to reach the size sufficient to make it a success because it requires the largest investment of time and money. It may take all your spare time unless you have help, and you may choose to build it into a full-time business. Of course, your success also hinges on the proximity of a processor and a laboratory, both of which will control the price you receive.

Absent from this list is anything with a lop ear. I like my rabbits to look like rabbits, not puppies. Pet buyers love the loppies, however, so if they are your market, then they are for you. Also absent are the giants and the dwarfs. I've raised them both and prefer those in between. Angoras are too much work for somebody with little time and no spinning wheel. If I couldn't have my favorites above, you *could* talk me into Satins, including the Mini Satins, or Standard Chinchillas.

Choose the Right Rabbit

Once you've identified your purpose, what remains is for you to pick the best breed for you. It's helpful when making that decision to be familiar with the many breeds. In the next few pages, I have categorized them by weight and fur type. Then I described the potential of each type for your consideration.

Remember, success in rabbit raising demands that you produce more rabbits than you are able to house. Even if you raise them for only the purest of motives — the joy of having them around — you simply will not be able to house them all. More practically, the producer of show stock must constantly make room for younger, improved animals. Those who produce breeding, meat, laboratory, and even pet stock must find buyers.

All of the various breeds fall into four categories by weight. You'll find photos of some breeds and all weight classes in this book (see the breed gallery on page 208).

The Giants — A Bit Too Big

Appropriately, the largest rabbits are called the giants. Among them are the Flemish Giant, the largest, which weighs up to 20 pounds (9.1 kg) and even more, and the Giant Chinchilla and the Checkered Giant, each of which weighs up to 16 pounds (7.3 kg). Some producers of meat rabbits raise the giant breeds, but most do not. They prefer those in the medium-weight range. Bigger is not necessarily better where meat is concerned. The giants' bones and appetites rival their size, so their feed-to-meat conversion ratio makes them less profitable than the medium breeds.

Nevertheless, some people like the way the giants look and enjoy showing them or selling them as pets.

Middleweights — Right on Target for Meat

Medium-weight rabbits reach 9 to 12 pounds (4.1 to 5.4 kg) at maturity. Among the most popular are the New Zealand Whites, Californians, Satins, Palominos, and Champagne D'Argents. Meat producers prefer this group because their ideal rabbit is a meaty, fine-boned fryer weighing 4 pounds (1.8 kg) or more at eight weeks of age. Medium-weight rabbits come closest to that target.

There's a Place for the Little Ones

Small rabbits include, among others, the Tan and the Dutch, the Florida White, the Mini Lop, the English Spot, and the Havana. These weigh 4 to 7 pounds (1.8 to 3.2 kg), are popular as exhibition animals and pets, and have considerable utility value either as meat for small households or in the laboratory.

Dwarfs — Popular as Pets and for Show

Barely nudging the scales at a mature weight of only 2 to 4 pounds (0.9 to 1.8 kg) are the Netherland Dwarf, the Polish, the Britannia Petite, the Mini Rex, the Mini Satin, the Jersey Wooly, the Dwarf Hotot, the Holland Lop, and the American Fuzzy Lop. These little guys displace hamsters, gerbils, guinea pigs, and even cats and dogs as pets. If you plan to raise pets, these breeds are ideal. If you want to show rabbits but have little space, they may fit into your plan.

Coat Types

Different rabbit breeds exhibit varied characteristics, coat type being an obvious one. Coat type can be classified as normal, rex, satin, or wool.

Normal Fur

Normal-furred rabbits dominate the species. You will find varieties in each weight group. The Flemish Giant, the New Zealand, the Tan, and the Netherland Dwarf all have this type of fur. Normal fur is about an inch (2.5 cm) long. When stroked toward the rabbit's head, good normal fur returns quickly to its natural position and lies smoothly over the body. The underfur is fine, soft, and dense.

Rex Fur

Rex fur is worn, naturally, by Rex rabbits, which are in the medium-weight group, and by the Mini Rex, among the smallest rabbits. Velvety and plushlike, rex fur stands upright and has guard hairs almost as short as the undercoat. Rex fur is only about ⅝ of an inch (1.6 cm) long.

Satin Fur

Extremely popular are the Satins, which come in a wide range of colors. They are found not only in the medium-weight but also in the small-weight class, as Mini Satins. Satin fur consists of a small-diameter hair shaft and a more transparent hair shell than is displayed by normal fur. This greater transparency of the outer hair shell gives the Satin fur more intense color and more sheen than normal-furred breeds, with the exceptions of the Tan and the Havana, which have coats that lie flatter than those of other breeds, thus presenting more luster and sheen. Satin fur is about an inch (2.5 cm) long, like normal fur.

Angora Wool

We can more properly describe the coat of the Angora rabbit as wool. Angoras, small- to medium-size rabbits, appear quite large because of their fluffy wool coats, which are 3 to 8 inches (7.6 to 20.3 cm) long. There is no mistaking Angoras; their coat is anything but subtle. You will know them when you see them.

3

Obtaining the Right Foundation Stock

BETTER THAN ANYTHING ELSE, purebred stock will help you start raising rabbits the right way, no matter why you raise them.

Purebred rabbits are pedigreed. Pedigreed rabbits may or may not be registered, and registered rabbits may or may not be worth buying, but nevertheless will be important to you. In this chapter, we'll discuss their relative significance. But first, consider crossbreds.

Crossbreds Are the Least Valuable

Crossbred rabbits are mongrels. They are of mixed ancestry, perhaps of the Heinz 57 variety or even something very close to purebred. They might even *be* purebred, but if they lack pedigree papers, they always will be considered crossbred.

Crossbreds carry the least value of any rabbits. You don't really know what ancestors came before them, and nobody really knows what progeny will descend from them. I have seen lots of crossbreds, and I wouldn't consider starting out with any of them. You can't predict what they will do for you, other than eat as much as or more than the best rabbits while returning to you the least satisfaction or money. For the most part, nobody wants crossbreds — not even meat processors, although they will probably take them. Processors can't predict how crossbreds will dress out: whether they will get lots of meat or lots of bone for their efforts. (There can be exceptions in the meat category, as mentioned earlier, because certain carefully calculated crosses or hybrids can produce successful meat rabbits.)

You don't know how much mongrels will cost to feed, how readily they will breed, how prolifically they will produce, how hardily they will resist disease, or anything else about them. If you like surprises, crossbreds are for you. If you want profitable, healthy, thrifty, handsome rabbits, forget about crossbreds. Ironically, the 4-H kids in my locale who insist upon keeping crossbreds enter these rabbits in shows in a class known as "commercial," and they call the crossbreds "commercials." The fact is that these so-called commercials have less commercial value than any other rabbits, particularly those "noncommercials," the purebreds.

A Pedigreed Rabbit Has Papers

A purebred rabbit is bred to a certain written physical description, called a *standard*. Purebreds are also called *standardbred* or *thoroughbred*, and these terms are nearly synonymous with another used to describe them — *pedigreed*. A pedigreed rabbit has a written record of its ancestry as evidence of its pure breeding.

Along with your purebred rabbit, you should get a pedigree paper, which may be handwritten, typed, or computer generated from an ancestor database. Note that the typical pedigree shown on page 28 contains the names of and other information about three generations of ancestors on both sides of the family. The pedigree also includes private ear identification numbers, registration numbers if registered, color, weight, any winnings, and perhaps a note or two about the size of the litter in which it was born, or how it ranked in the litter relative to the others, in the opinion of the breeder. The more information, the easier it is to determine the rabbit's ultimate value as a breeder of other good rabbits.

No Paper? Not Pedigreed

Remember this about pedigreed rabbits: If you don't get the pedigree paper, the rabbit is not, for all practical purposes, pedigreed, no matter how purebred it actually may be, because you can't prove it. Without the

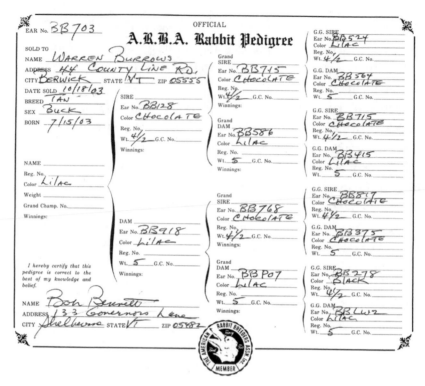

A typical pedigree paper

paper, the rabbit is worth considerably less as a breeder. If you are wise, you will not take rabbits without pedigree papers.

Registration

A pedigreed rabbit may also be a *registered* rabbit, but the two terms are not synonymous, although many newcomers to rabbit raising appear to think so. To gain registration, a rabbit must be purebred, have a pedigree paper, and be a mature specimen. It must also be examined by a licensed American Rabbit Breeders Association (ARBA) registrar, who must certify it to be free of apparent physical defects. In the registrar's opinion, the rabbit must meet the *minimum* physical requirements of the breed as described in the written breed standard. It should, for example, be of the correct size and weight, color, and body type. The registrar must certify, by signing an affidavit, that the animal passed the inspection. A registrar's reputation and right to the registrar license are put on the line with every such signature.

Remember, with rabbits there is none of the American Kennel Club sort of thing, where you simply send in the pedigree papers of the sire and dam and your litter gains registration. A physical examination is required. This procedure makes rabbit registration just about the best animal-registry system there is. Only members of the ARBA may apply for registration of their stock, but anyone may, of course, buy and own such animals. The small fee for registration is paid to the registrar, who

The registration certificate of the winner of best of breed at a national ARBA convention show keeps some of the money and sends the remainder to the ARBA, which files a copy of the registration application at its headquarters and, if the application is approved, issues a certificate to the owner.

What Does It All Mean?

Now what does all this discussion of papers really mean to you, the new breeder? Why should you care if rabbits you are going to raise only to sell as pets or laboratory animals or simply to eat have a pedigree or registration certificate behind them?

Well, it not only matters, but actually it is vital if you are to start right and be a success. That's what this book is all about, and if it could have only one chapter, this would be it.

Papers Are the Proof

Let's suppose you have decided to raise meat rabbits for your own table. You have calculated that the low feed cost to produce a pound of tasty meat for your family is worth the effort you will have to make to put it on the table. And you have chosen New Zealand Whites as the best breed to deliver the goods for you. So far, so good.

Now you go out to buy New Zealand Whites. You see a sign along a rural road, RABBITS FOR SALE. The owner shows you white rabbits and says they are New Zealands. But how do you really know that's what they are? You don't, unless the owner shows you some records of ancestry — pedigrees — to prove his claim. The owner knows he must keep these records and use them to maintain a herd of breeders that meets the standard for New Zealands.

Understanding the Standard

Let's go back a bit and discuss the standard. This written physical description of the breed is prepared by members of the breed specialty club — in this case, the American Federation of New Zealand Rabbit Breeders. A *standard committee* of members, all expert breeders, decides precisely what the New Zealand White will look like. It produces a written description. Members of the federation vote on it. If a majority agrees, the standard is submitted to the ARBA, which is the parent organization of all the breed specialty clubs. The ARBA Standards Committee, if it votes its approval of the standard, admits it for publication in a volume entitled the *Standard of Perfection*, which the ARBA revises and publishes every five years.

The *Standard of Perfection* includes the standard for every breed that the ARBA recognizes. Every raiser should own this book, or at least a copy of the standard for the breed being raised. You should study the standard to make certain you are carrying out a breeding program that will produce animals in accordance with it. For example, in the case of a New Zealand, if the fur is woolly, the rabbit will be disqualified in a show and may produce woolly offspring. Not only would that soon lead to a rabbitry full of disqualified rabbits, but the fur would be worthless. That makes a difference when you breed for show, certainly, but also if you attempt to sell breeding stock or even meat to a processor who sells the pelt. Nobody wants disqualified stock.

You May Have Found Good Rabbits

If the breeder with the sign along the road is familiar with the standard, has breeding records of all stock, and provides a detailed pedigree, you probably have found good stock. If the rabbits are also registered, you know that they have met the minimum physical requirements for the breed, and you can view the stock with great confidence. If, on top of that, the pedigree and registration papers indicate show winnings, the rabbitry is clean, and the stock is obviously well cared for and prolific (that is, with lots of litters on the grow), you have found a breeder from whom you certainly might buy.

Several Assurances

For you, the prospective producer of meat for your table, there are several assurances here that you otherwise would not obtain. These white rabbits, purported to be New Zealands, will in fact reproduce their own kind, will gain weight properly, will convert feed to meat efficiently, will produce a usable or salable pelt, will be healthy if properly cared for, will be prolific, and will otherwise be attractive and desired by others. They will, in fact, be worth buying. They will, in fact, start you out raising rabbits right.

Now you might ask if the owner could have forged all this paperwork, and the answer is yes. It has been tried before, but you're not likely to run into this problem. A pedigree is only as good as the person who writes it, to be sure. But if the person bothers to write it at all, it's likely that records are kept. Keeping track of records suggests a pretty good job of breeding. The raiser who keeps no breeding records at all has nothing to offer you.

The Best Rabbit Raisers

With all this in mind, let's take a look at the persons best qualified to sell rabbits to you — the ones who will start you right.

Buy rabbits from only the very best, most highly respected raisers of your chosen breed that you can find. There are probably only a handful of these people for the more obscure varieties, but the United States boasts thousands who raise the most popular breeds. There is good breeding stock to be had, and it isn't that difficult to find.

The person from whom you should buy is one who, if a producer of meat and laboratory stock, consistently out-produces competitors and gets top dollar for production. Processors pay a premium. Laboratories seek out this raiser. Such a person prospers from rabbits, far beyond "just paying for the grain."

A Winner — With Registered Breeders

If the rabbit raiser you're considering sells show and breeding stock, you want a consistent winner, with trophies and ribbons to show for it. Lots of them. Not the top prize every time — nobody does that — but up there among the winners almost every time.

In addition, the raiser you want keeps breeding stock registered, providing a pedigree paper for every rabbit. The raiser produces quite a few rabbits — not necessarily thousands but probably hundreds every year. The raiser is an active member of the ARBA, of breed-specialty clubs, and often of local clubs. Such raisers show their rabbits at almost every opportunity. Their reputations, within at least the breed-specialty club, are national. There are plenty of satisfied buyers to talk to about these reputable individuals.

Satisfaction Guaranteed

Excellent raisers stand behind every rabbit they sell. Their rabbitries are neat and clean, and their rabbits are clean and perfectly healthy. These raisers are delighted to show the rabbits to you. They want you to see the fine conditions under which the rabbits are kept. They guarantee your complete satisfaction and will refund your money or exchange your rabbits for any reason, even an unreasonable one.

These raisers make such guarantees because they raise rabbits primarily for the satisfaction derived from the activity. If you aren't happy, they aren't happy. But these raisers won't let you pick out the rabbits, at least not on sight alone. Only the raiser knows, for example, which buck and doe to pair up for breeding success. That's not something you can possibly know without being familiar with their ancestry.

How to Find a Raiser

How will you find the rabbitry that will supply your foundation stock? Probably *not* by a sign along the road. First, decide on a breed.

Select a Breed

Follow the recommendations in chapter 2. If you haven't yet made up your mind, attend a nearby rabbit show where most breeds will be on display. Look them all over. You can find the date and location of the nearest upcoming show in *Domestic Rabbits* magazine, which you will receive when you join the ARBA. Write, e-mail, or call the listed show secretary if you need directions to the show. Make sure you see as many different breeds as you possibly can. By doing your homework this way, you will be much happier with your eventual choice and less likely to change to another breed in short order.

When you've decided on a breed, watch it being judged and talk to the exhibitors. Ask where they got their breeding stock — and who *they* recommend. Perhaps they will have some stock for sale themselves, but don't buy any at the show. It's a poor idea to buy rabbits in a showroom, as I explain below, especially if you want to buy them just because they are winners.

Resist the Temptation to Buy at a Show

I see many people purchasing rabbits at shows, and I sometimes sell them there myself, but I really don't think too much of the idea for beginners. It's all right for the experienced exhibitor, who knows the reputation of the seller and perhaps adds a buck or doe to the herd in an effort to improve certain characteristics or simply to infuse new blood into the line.

But for you, the beginner, the show is really an information-gathering excursion. You might become interested in the rabbits of a certain exhibitor. If so, obtain an invitation to visit that rabbitry. If it meets the criteria just covered, it may be a good one from which to buy.

Inquire at Area Feed Stores

Another way to find good rabbits locally is to inquire at area feed stores. All rabbit raisers go there regularly. The store personnel know who they are, obviously, because of the feed they buy. So make some phone calls and visit them. You can also post a 3×5 card "want ad" on the store bulletin board (and later, you can make some sales the same way, by posting a 3×5 card "for-sale ad").

Mail-Order May Be the Answer

If you can't find suitable raisers locally, you always can find them elsewhere. Consider purchasing your breeding stock by mail. Check the classified pages of *Domestic Rabbits* magazine. You'll find names of raisers who have stock for sale. These people can supply stock to newcomers and old-timers alike all over the United States, and their rabbits are so respected that they are bought sight unseen.

Join a Breed-Specialty Club

Before you buy, determine who has the best stock to sell. Join the breed-specialty club of your chosen breed, and read the club newsletter. In the newsletter you'll find show reports and sweepstakes point standings, and it is from among the owners of the winners of these shows, and leaders of these standings, that you will want to consider making your purchase. If these members are among those advertising stock for sale in *Domestic Rabbits* or in the club newsletter, write and tell them what you propose to do. You can do so with great confidence that your letters will be well received and that you are on your way to obtaining good stock. You are writing to raisers who are among the best in the nation.

Surfing for Rabbits

More and more rabbit raisers are using the Internet to communicate, so you might also check out Web sites and then e-mail respected raisers to inquire about their stock. The ARBA and many breed specialty clubs near and far maintain Web sites that include names and addresses of members and information about the breeds. You are likely to find some of them in your vicinity. You can access information by simply Googling the name of your chosen breed. From there you can do a lot of research that may well prove fruitful.

When you become an established breeder and have rabbits for sale, you may want join a breed specialty club and take part in Internet transactions. You can pursue the Internet further simply by Googling "rabbit" to get other leads with Internet trading posts and classified ad sites. Just remember: When you find what looks good, apply the criteria discussed on these pages.

The Right Raiser for You

First of all, those who will start you out right began with the best possible stock themselves and have consistently improved it over a long period of time. They have shown and won consistently, and though they breed a large number of rabbits, they sell only the best as breeding stock. Many of their animals go for meat or to laboratories, even if the raisers have been at it for some time, because their standards are so high and get higher every year. The rabbits they would have sold or kept for breeding stock last year will not measure up to this year's output, because they've worked hard to upgrade their herds through selective breeding — a never-ending process. Also, the best raisers charge a good price for their breeding stock, so they can afford to advertise. And they advertise so they will be able to breed large numbers of rabbits continuously and thus stay ahead of other breeders. All of the above is precisely why their rabbits are some of the best of the breed that you can find.

Such raisers will sell you good stock and will stand behind it. They will guarantee that these rabbits will do well for you if you do right by them. Such people will happily answer all your questions and be interested in your success, which of course is an extension of their own. They have a vested interest in you: They want you to make good with their stock. Later on, they hope to read the show reports in the same newsletter where you first read about them and find that you are now among the winners in your area's shows. You can be sure they will derive a lot of satisfaction and one of their chief rewards from your success.

What to Expect When You Buy

Now what kind of rabbits should you expect to be able to buy? You should not expect to buy anyone's very best rabbits — not at a reasonable price, anyway. But you should expect young stock out of the best or some of the older breeders, sold to make room for promising youngsters. Don't expect grand champions, unless you want them much more than the owner and are willing to prove it with your pocketbook. Don't even expect to buy show winners, although the young stock you purchase may in fact go on to win great titles. Buying a show winner proves only, in most cases, that you can spend a lot of money. Producing your own show winner is light-years more satisfying. It is fair to add, however, that some people produce so many winners at so many shows that they do have winners for sale at reasonable prices. If such is the case, do not hesitate to take them.

Four Is a Good Number for Starters

Any successful raiser will tell you to start your rabbitry small. Even if you plan a large rabbitry eventually, don't expect to purchase the whole herd at once. Start with a minimum of four rabbits. I wouldn't try to build a

---

RABBIT AGE

A *junior* rabbit is younger than 6 months; a *senior* rabbit is 6 months of age or older (8 months for the medium and giant breeds).

herd from a single pair or even from a trio of two does and a buck, which is a popular way to start. A couple of junior bucks, a junior doe, and a bred senior doe make an ideal start.

If you ask for two junior bucks and a junior doe, you should expect to get them from the very best of the top breeders in the herd. The senior doe might be as old as 2 years and may have only a year or so of breeding life left. Nevertheless, if this doe has been producing in a good rabbitry for 18 months already, she must be pretty good to have hung around for so long.

Ask to have the senior doe bred to the best buck the owner recommends. By the time she's ready to mate again, the junior bucks will be mature. Also, the junior doe will be ready for mating. So in 3 or 4 months, you could have your original four rabbits, a litter of youngsters, and two more litters on the way.

Starting out with two bucks and two does means that you will always have the option of selling some breeding stock yourself, no matter what your reasons for wanting to raise rabbits. You want a fairly broad base for further breeding. With two bucks and two does, you will be able to sell and breed pairs and trios from two different litters from four different parents. Such offspring usually are more desirable than, for example, a brother-sister pair. See chapter 6 for more on breeding, but for now note that two pairs are the minimum.

Two does and two bucks make an ideal start.

THE BEST RABBITS EAT NO MORE THAN THE WORST

You say you want rabbits only for your own table, so why bother to invest in the best? Again, I say it costs no more to feed and house the best rabbits, and your potential is much greater, even for producing meat for your own table consistently, if you start with the best breeding stock available. Also, you will take a lot of pride in good rabbits and none in bad. Furthermore, they will be admired and sought by others, so you will find a demand for them. Keep in mind that you will always, if you are a success, produce more rabbits than you will be able to use yourself. Good as it is, nobody wants to eat rabbit *every* day.

Rely on the Raiser's Judgment

Depend upon the raiser's judgment as to which pairs to buy. Don't insist upon unrelated rabbits. In chapter 6 I'll cover various methods of breeding, but for now let me say that more often than not, unrelated rabbits will not do the job nearly as well as related stock. Your main task is to select and put your trust in a raiser of good reputation, whether you purchase your rabbits by mail-order or in person. Have the owner select your stock for you. Nobody knows the stock as well as its owner, and nobody knows better how to select which animal will go best with another.

How to Build Your Herd

Let's suppose you take my recommendation and purchase four rabbits in January. The senior doe *kindles*, or gives birth to a litter, in February, and you decide to save the best three does and the best buck from this litter. Later on, in May, both the junior doe and the senior doe kindle. You save the best three does and the best buck from each of these litters. By summer you have bred the original senior doe again, the original junior doe, and the three does from the first litter. You have five litters on the way, but you also have six junior does to breed in the fall and a choice of four eager young bucks with which to mate them. Before winter you will have eleven breeding does, more juniors growing, a fine selection of stud bucks, and litters all over the place. Your four original rabbits are now one hundred of only your best.

Maybe you don't want one hundred, and maybe you won't get quite that many, but it could happen if you want it to. It is a simple matter to calculate just how fast you can reasonably start from the above example. Remember that you'll need to provide housing for all these rabbits, so available space will have a bearing on your start. You need not, of course, save as many young does as I'm suggesting here. If you do not want as many rabbits, by all means save only the very best doe from each litter and replace one of your original does with a better one, should it come along. But please do not start with fewer than four rabbits if you have the slightest intention of selling any breeding stock or of building yourself even the most limited herd.

When making choices about building your herd, keep in mind that you — and your breeding-stock buyers — will have a greater demand for does because does have a shorter breeding career than bucks. The average doe bears litters for about 3 years. Bucks will sire litters for 6 to 9 years.

What You Can Expect to Pay for Purebreds

I correspond with many rabbit raisers regularly, and for years I have asked them how much they charge for the purebred stock with pedigree papers that they raise. The following chart gives the averages of these prices. It should be stressed that the prices are only averages, and that actual prices can vary greatly. Prizewinning animals may sell for a lot more. But the chart can provide you with guidelines: You shouldn't pay much more than these prices unless you have evidence that the individual animals are spectacular. On the other hand, I would be wary of a $5 price tag too.

Purchasing by Mail-Order

It is a simple matter to purchase rabbits by mail-order. First, have your hutches ready. Then send the raiser a check or money order for the rabbits, along with a letter giving your complete name, mailing address, and phone number. The two of you agree on a shipping date. You will be

AVERAGE BREEDING STOCK PRICES

Breed	Junior Average $ Value	Senior Average $ Value	Breed	Junior Average $ Value	Senior Average $ Value
American Fuzzy Lop	33	62	Havana	15	25
Angora (English)	50	60	Holland Lop	38	51
Angora (French)	27	51	Hotot	50	50
Angora (Giant)	90	115	Jersey Wooly	26	56
Angora (Satin)	35	75	Lilac	25	35
Britannia Petite	73	73	Mini Lop	19	41
Californian	10	25	Mini Rex	22	32
Champagne D'Argent	10	25	Mini Satin	30	80
Checkered Giant	15	20	Netherland Dwarf	37	54
Chinchilla (American)	10	25	New Zealand Black	15	30
Chinchilla (Standard)	15	43	New Zealand Red	34	65
Dutch	25	50	New Zealand White	25	45
Dwarf Hotot	30	58	Palomino	32	32
English Spot	29	48	Polish	25	40
English Lop	35	87	Rex	23	40
Flemish Giant	27	70	Rhinelander	25	40
Florida White	35	60	Sable	23	33
French Lop	16	29	Satin	25	33
Harlequin	15	28	Silver Fox	25	45
			Silver Marten	18	33
			Tan	41	64

Overall Junior Average: $30; Overall Senior Average: $55

Note: This survey, which I conducted annually for about 15 years, included an average price of $16 received for pet rabbits of all breeds.

phoned by the airline or air-express office at the airport nearest you to pick up your rabbits. The shipping charges will vary depending on the number and weight of the rabbits and will be most economical if they stay on one airline during the trip. There will be an additional charge for veterinary health certificates, which are required by the airlines.

The seller can get your new rabbits to you the same day if they are shipped from and received at major airports. If both of you live in Podunk, it could take another day. Fear not for the comfort of your rabbits, for they are hardy travelers and are shipped with sufficient food and water for their journey. In my own case, living near Burlington, Vermont, I am able to have rabbits received all over the United States the same day I ship, about the same time it takes rabbits to travel to and from a show. Same-day shipping helps get the rabbits to their destination in good condition. In fact, to my knowledge, of the many hundreds I have shipped, only one rabbit has died en route. Of course, the airline made good on it because I insured the shipment. Everything else being equal, a seller who ships from a major airport increases your chances of a successful trip for your rabbits. Rabbits may be shipped only by air — not by mail or package delivery service. In some cases you may be able to find a local independent trucker willing to move them.

Bringing Your Rabbits Home

When you get your rabbits home, install them in their hutches with food and drink, and then leave them alone. They need a little time to get used to their surroundings. Write or phone the seller to say you have received them, that they are okay, and that you are a little thrilled with them, which will be the case, of course, because you've taken all my preceding advice. When you have officially accepted the rabbits, the seller will send you the pedigree papers, each one marked with a number that corresponds with the number in each rabbit's ear. You should also get the date your senior doe was bred and a copy of the pedigree paper for the buck to which she was mated.

On the off chance that you are not 100 percent pleased with your purchase, you have every right to demand another set of rabbits or your money back — and I would expect to get every cent except perhaps the shipping cost. A good seller will be happy to replace the rabbits to make sure his or her good name is not besmirched. It is in the everlasting best interest of sellers to ship excellent specimens.

HUTCHES FOR EACH ADULT

Each buck should have his own hutch, and it's best to have a hutch for each young doe, although two does to a hutch is a satisfactory arrangement until mating time. See chapter 4 for more on the housing needs of rabbits.

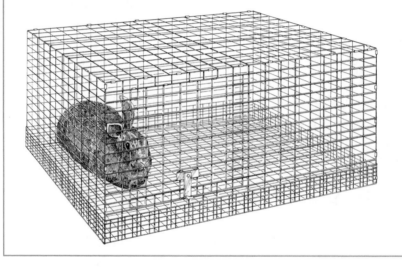

PART II

Housing, Feeding, and Breeding

4

Housing and Equipment

YOU WANT TO RAISE RABBITS. Plural. Multiple. As in rabbits. Caring for a single pet requires little space and forethought. Once you become a rabbit *raiser*, however, litters arrive. You need more space. You need a rabbitry.

You Are in Charge

Let's examine your needs, the needs of the rabbits, and the needs of the various constituents or stakeholders who can influence the enterprise. They include family members, neighbors, passersby, customers, and perhaps even the local government. Taking time to consider all the angles will go a long way toward creating a successful operation. It's all up to you: you are in charge. Next to the right foundation breeding stock, the right housing and equipment will do the most to ensure your success with rabbits.

Envision yourself as the president and chief executive officer of your rabbitry. Your first responsibility is the same as it is for the leader of the country: security. You must provide a secure rabbitry — one that protects your animals from disease, parasites, predatory wild and domestic animals, and even human threats to well-being and success. You must also safeguard them from each other, which means individual living arrangements for some and group homes for others. For example, each mature male will need its own space, to prevent fighting and possible injury. Young growing does can live in harmony as a group for the early months of their lives.

To start the process, think about space requirements, and think *inside* the box or wire hutch, the basic unit of a successful rabbitry. How large should each wire hutch be, and how many will you need? How many should you want? How will you maintain a rabbit population consistent with the space you provide?

Why the All-Wire Hutch?

Years ago in Europe and in North America, most domestic rabbits ran loose in a barn built primarily for cattle, sheep, hogs, goats, horses, or poultry — or all of them. (Of course, even earlier they coexisted with their animal pals on the first floor of the house!) In that domicile one didn't have to provide any accommodations for rabbits because they burrowed into the dirt floors or under hay or straw and formed a colony or warren. Feeding and watering was no problem. The farmer didn't bother. The rabbits foraged for themselves, snitching nourishment from the other species.

That was the good news.

The downside of this colony or warren situation, however, sank in more deeply the more the farmer thought about it (although he probably didn't). First, if he wanted a rabbit for dinner, he had to catch it. He wouldn't find out if he had captured a tender fryer or an ancient stewer until he took a bite. Second, he couldn't depend upon a set quantity or output or the timing of production because the rabbits made that determination themselves. There was worse: Parasitic worms penetrated the animals from the dirt floors they shared with other species. That weakened the rabbits and left them thin, unthrifty consumers of the available feed and susceptible to debilitating or even fatal disease.

Foxes, weasels, and other four-legged predators devoured some. Owls and hawks swooped down on others. Mature males often fought and injured each other, and some survivors hopped about outdoors, only to provide more meals for wildlife or household cats and dogs. In addition, because rabbits love to gnaw, they chewed portions of the barn. If you think that barn qualified as confinement, think again, because what they really had was a Swiss cheese prison supervised by inmates and enemies but not by the warden.

Later, a somewhat better way to keep rabbits arrived with the Morant hutch—a portable, floorless, outdoor enclosure occupied by a single rabbit or several of both sexes that was moved around from place to place by the farmer. The rabbits mowed the grass and fertilized it too, but because there was no floor, the parasites flourished, and sometimes the rabbits burrowed under and out to an uncertain fate. Grass provided mere subsistence, although it was sometimes supplemented by vegetable garden waste. The quality of the livestock and of their offspring was irregular. Mortality rates were high.

A recent popular notion, held mostly by city dwellers but not by experienced farmers and other savvy country folks, proposes that poultry should run on the ground amid their own droppings as they did in days of yore, eating the bugs and those worms that don't eat them (and requiring chemical worming). Parasite infestation was the plight of rabbits housed in Morants and of animals from barn-housed warrens or colonies. These points make the case for the all-wire hutch, an enclosure that houses rabbits securely and under the cleanest conditions of any livestock. Your prized rabbits never sit in manure for even a minute and don't contract parasitic worms. Also, you control the diet, the selection and timing of matings, and herd health and size.

You may hear from the well-meaning, but naive and uninitiated, that any confinement housing really only robs the rabbits of freedom. The fact is, however, that well-bred domestic rabbits have been raised in wire hutches for thousands of (rabbit) generations and have never experienced freedom. They don't miss it. You can't deprive them of something they have never had. In fact, if one of your rabbits should escape, it wouldn't know where to turn, even if it wanted to flee. Should the escape occur in your absence, the vulnerable animal, lacking the fear-inspired instincts of its wild cousins, would become an easy mark for passing predators. You should know that some types of hutches are virtually escape proof, which I recommend.

CRITERIA FOR THE BASIC HUTCH

Consider first what kind of living conditions rabbits require. They need clean quarters, plenty of light and ventilation, and protection from winds and drafts, although they can withstand plenty of cold. They cannot tolerate extreme heat, however. They need a dry hutch, all the way around: no leaky roof; no wet floor; no dampness. They don't need a lot of room. Domestic rabbits are not used to hopping great distances like their wild cousins. But they do need enough room to move and to rear a family comfortably without overcrowding.

Next, what about you, the rabbit keeper? What kind of hutch do *you* need? The hutch must prevent your rabbits from escaping. It must be easy to clean, self-cleaning if possible. It must be easy to handle and be adaptable to inside and outside use (you may start out with outside hutches and later move them into a shed or other building). It must be durable. It must allow you to feed and water the rabbits conveniently, without even opening the door, if possible. It must allow you to see what is going on inside at all times, because it is the observant rabbit keeper whose stock thrives. It must allow you to catch and handle the rabbits with ease. And it must be inexpensive and easy to build or buy.

The wire hutch is, in fact, all wire. No wood. Too often a prospective rabbit raiser obtains lumber and chicken wire, hammer and nails and begins to build. Others buy expensive wooden hutches from a pet store or online. Amazingly, there are lots of them available, and some are extremely expensive, besides being ridiculously unsuitable. Rabbits gnaw the wooden frame. They soak it with urine. Because wood framing members support the wood or wire floor, droppings pile up in the corners, despite daily cleaning. A wooden hutch is damp and smelly and a breeding ground for parasites and disease.

There is only one kind of hutch worth using, whether you build it yourself or buy it.

It is the most advanced hutch in use today and has proven its worth for the more than 60 years that welded wire mesh has been available. You can see from the illustration below that the hutch is self-cleaning. It requires only an occasional wire brushing to remove shedding hair, plus periodic disinfecting (something that's also needed but actually impossible with a wooden hutch). Droppings and urine fall right through to the ground or to dropping pans. Complete ventilation, so terribly important, is afforded. And the hutch remains dry at all times.

If you plan an outdoor rabbitry, the rabbits and hutch will need protection from the weather, depending upon your climate, but indoors the hutch is fine the way it is. Self-feeders steal none of the hutch floor space and can be filled from the outside. The most modern, labor-saving devices for water can be used, or crocks or bottles can be filled without opening the hutch. The more rabbits you have, the more time this saves. The hutch will last for many years, because the rabbits will not be able to gnaw it. And years from now, your wire hutch will still be serviceable and doubtless worth more money than it cost. It's a fact. I built some all-wire hutches 35 years ago. They are still just as good as new, although not as shiny. Today, as then, they cost less than any other kind.

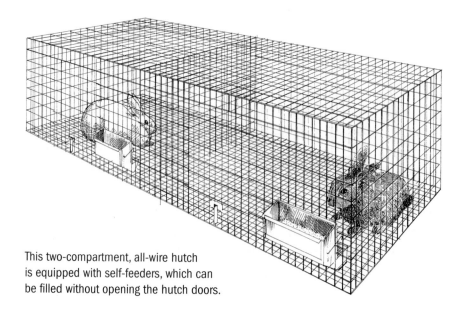

This two-compartment, all-wire hutch is equipped with self-feeders, which can be filled without opening the hutch doors.

The wire-mesh floors are perfectly suitable for rabbits' feet. Some people honestly but mistakenly believe the wire floor will hurt a rabbit's feet. They forget about Thumper in Felix Salten's book *Bambi*. Thumper was the rabbit that periodically stomped a thickly padded hind foot, or *hock*. Like Thumper, the domestic rabbit has very big, furry feet. There was also Joel Chandler Harris's Br'er Rabbit, whose favorite refuge was the thorny and prickly briar patch. Rabbits' feet are more than just lucky. They are perfectly designed for perforated floors. In fact, many rabbits on solid floors develop sore hocks, because wooden floors are impossible to keep clean and dry.

WELDED-WIRE ECONOMICS

The economics of buying cage wire and building your own hutches are changing. The cost of wire by the roll often makes it more economical to buy knockdown, prefabricated hutches that you merely assemble. The manufacturers of the prefabricated hutches can buy wire by the truckload or even by the freight-car load, so their costs for both wire and freight are much lower than what the individual must pay.

Another part of the equation is the lack of waste when you go the prefabricated route. When you buy 100-foot rolls, you always have more floor wire than you need for the amount of side and top wire you get from rolls of the same length. Yes, you can do the math and buy more of the 1 × 2-inch wire for sides and tops, but despite my best efforts over the years, I always have some extra floor wire that I paid for but that just takes up space in the barn.

It really comes down to what you have to pay for the rolls versus the cost of the prefabricated hutches, so I encourage you to shop around and do some comparisons. Your farm or feed store probably carries both wire rolls and prefabricated hutches. You may also contact me by mail or e-mail for supplier recommendations. You will find contact information in the back of the book.

How to Make a Wire Hutch

You can build your first wire hutch just as well and as easily as I built mine. I have put together many hundreds of them and can now complete one in less than half an hour. But your first attempt will probably take a couple of hours. You won't make a mess of sawdust, nor will there be any loud banging or sawing. I build my hutches in the basement in cold weather and in my garage when the weather's nice. You will need only pliers and wire cutters, and your cost will be less than if you used lumber and screws and hinges and assorted hardware. The material quantities given here are for a wire hutch with a floor area of 2½ by 3 feet, fine for all but the giant breeds, which need an extra foot or two.

MATERIALS

For sides

- 1   length of 1 × 2-inch 14-gauge welded galvanized wire
       fencing (sometimes called turkey wire),
       18 inches wide by 11 feet long

For floor

- 1   piece of ½ × 1-inch 14- or 16-gauge welded galvanized
       wire mesh (not hardware cloth), 30 × 36 inches (14 gauge
       is heavier but more expensive)

For roof

- 1   piece of 1 × 2-inch 14-gauge galvanized wire mesh,
       30 × 36 inches

For door

- 1   piece of 1 × 2-inch 14-gauge welded galvanized wire mesh,
       12 × 13 inches
- Latch, dog-leash snap fastener, or wire coat hanger

EQUIPMENT

- Measuring tape
- Hammer
- 2 × 4 (2 feet long)
- J clips or C rings (about 80)
- J-clip or C-ring pliers
- Wire cutters

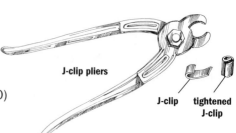

J-clip pliers

J-clip   tightened
J-clip

Prepare the Sides

1. Lay the full side piece of wire on the floor. (Don't cut it.)

2. Using a hammer, bend each corner around the length of the 2 × 4 to make the two 2½-foot and the two 3-foot sides. Don't bend against the welds.

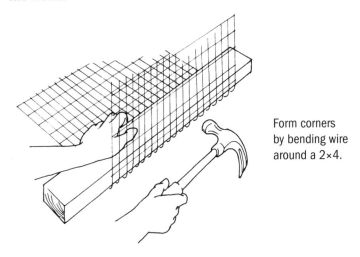

Form corners
by bending wire
around a 2×4.

3. Fasten into a rectangle by clamping the J clips or C rings on the wire with pliers about every 3 inches or so. If possible, use J-clip pliers. You now have all four sides of the hutch assembled.

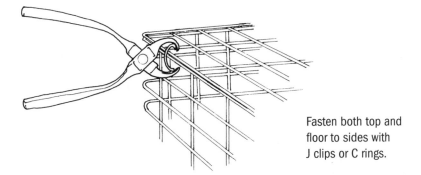

Fasten both top and
floor to sides with
J clips or C rings.

Prepare the Floor and Roof

1. Fasten the ½ × 1-inch floor mesh piece to the sides to make the bottom, using rings or clips and pliers.

2. Fasten 1 × 2-inch roof mesh piece the same way to make the roof.

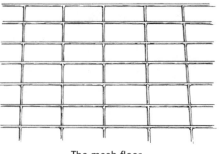

The mesh floor

Prepare the Door

1. Using the wire cutters, cut a door opening 1 foot square on one wide side of the door piece. This will be the front. Leave ½-inch stubs.

2. Bend the stubs back with the pliers so there are no sharp edges.

3. The wire piece for the door is 13 inches square because the door should overlap at least ½ inch all the way around. Many rabbit raisers prefer a door hinged at the top (with J clips or C rings) that swings up and into the cage. That's how I build mine and the way commercial rabbitries want theirs built. This way the door is inside the hutch when open and not extending out into the aisle, where it catches sleeves or otherwise gets in the way. Another good thing about a top-hinged door is that if you forget to latch it, it is still closed and prevents escape.

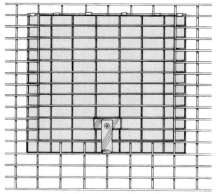

Door overlaps opening, except at top, where it is fastened with J clips or C rings. Fasten door to inside of hutch, so it swings up and in.

Galvanized metal door latches swivel left and right to open and shut.

Creature Comforts: Situating the Hutch

If you plan to use this hutch indoors, perhaps in a garage or shed, you may want to put a pan or wooden box lined with plastic sheeting underneath it to catch the droppings. Outdoors, the hutch will need a roof and sides. You might want to build the legs and roof together in table fashion and hang one or two tiers of hutches with hooks from the roof. The legs should be high enough for the hutch to hang at a level that's convenient for you to feed and water the rabbits but also above any possible splashback from rain and deep snow.

Use plywood and 2×4s, especially pressure-treated 2×4s (see illustration on page 54 for construction design). On cold or rainy days, hang a heavy plastic or canvas tarp from hooks mounted on the front. Surround the shed with a sturdy fence to protect rabbits from predators.

If your hutches are outside, make sure your rabbits are out of the wind in winter and in the shade in summer. Try to locate hutches in a shady plot protected by a high, sturdy fence. Dogs can be a problem if they roam free, particularly in packs. All-wire hutches can also be suspended from a strong chain-link or board fence, eliminating the need for legs and perhaps even back-wall weather protection.

No matter how hutches are placed, make every effort to supply shade. A double roof, with 3 or 4 inches between layers, is good protection from the sun if you don't have shade trees. Some raisers grow squash or pumpkin vines over their hutches to provide shade. If you have space available in a well-ventilated shed or garage, your rabbits will be safe, and both you and they will be more comfortable in every possible kind of weather.

HUTCH SIZE FOR DOES AND THEIR LITTERS

A doe raising a litter needs a hutch with square footage nearly that of her adult body weight. For example, I have 5-pound does in hutches that are 2½ by 2 feet, or 5 square feet. This rule of thumb is for rabbits whose litters are weaned at eight weeks. Some raisers wean them earlier and thus use a smaller hutch.

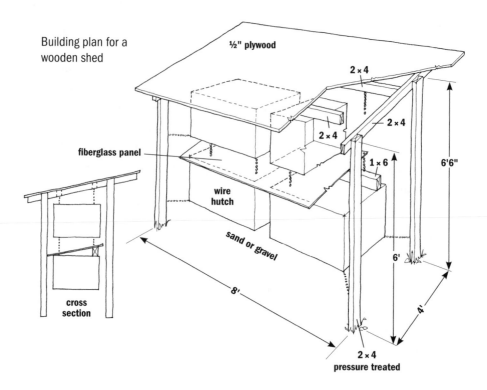

Building plan for a wooden shed

½" plywood

2 × 4

2 × 4

2 × 4

1 × 6

6'6"

fiberglass panel

wire hutch

sand or gravel

6'

cross section

8'

4'

2 × 4 pressure treated

You may also want to consider a shed of some sort to keep your rabbits out of sight, if you have neighbors who might be concerned. A shed or at least a fence keeps the rabbits out of sight and out of mind. A hedge also works, and so does a stack of firewood. One rabbit raiser stacked firewood like a fortress, in a rectangle. He installed his hutches at night inside it. His neighbors never saw any rabbits or hutches. Later, he planted a hedge around the perimeter. A few years later, when the hedge grew tall, he removed the firewood.

If you do put your hutches inside a building, they are best placed over a sand or gravel floor, which absorbs moisture. A barn is great and may have such a floor, but some breeders keep their rabbits in other kinds of outbuildings, even the garage. My own began in a steel shed specially constructed from two prefabricated steel storage buildings that I modified to provide extra ventilation. At present my rabbits are in a barn that I built myself and that I will describe later in this chapter.

THINK OUTSIDE THE HUTCH

Before you choose a location for your rabbitry, you need to determine what, if any, external forces could affect your success. If you live on large rural acreage, you can pretty much do whatever you want. Of course, lots of people raise rabbits because they like livestock but don't have access to agricultural land. If they did, they might choose larger animals. A great thing about rabbits is that they require little land, make nary a peep nor any other sound, and can be raised under attractive and nearly odorless conditions.

Nevertheless, your family members and neighbors must be considered. Your municipality may have zoning restrictions that you should know about. Check on the law first, but don't march into the town clerk's office and ask, "Is it okay to raise rabbits in this town?" Instead, ask for a copy of the zoning ordinances. You might or might not be able to take home a copy, but assuredly you may peruse the regulations in the town office. Depending on what you find out, you can decide what to do next.

In many municipalities there is no law on the books specifically about rabbits. In some, laws may be ambiguous, not mentioning rabbits by name but merely discussing "farm animals." In others, there may be ordinances easily circumvented. The fact remains, however, that even where there are no rules against raising rabbits, if your rabbitry doesn't pass the sniff test of family and neighbors, it won't get off the ground, or if it does, it won't last long. An attractive or unobtrusive rabbitry stands the best chance of success. A rabbitry that's an eyesore can, in the opinion of critics, "smell bad" or even "make neighbors sick," *regardless* of legality or actual sanitary conditions.

Recommended Hutch Placement within a Rabbitry

In regular correspondence with many rabbit breeders, I learn how they arrange their hutches inside rabbitry buildings, whether barns, garages, or sheds. Here are four points to consider:

1. **Don't put hutches too close to the walls of the building.** If you place your wire hutches against walls, or only inches away, those walls soon will be sprayed with urine, coated with molted hair, and even caked with manure. Besides the resulting unpleasant smell and breeding ground for flies and disease, your wooden walls will rot or your metal walls will rust. Give yourself at least a 3-foot aisle between walls and hutches.

This slant-roof shed was built from angle iron and plywood. The wire hutches are suspended within from S hooks and are removable. Corrugated fiberglass roofing was used for sloping droppings boards above the lower tier. Note the plentiful shade this location provides. See directions for building on pages 68–69.

2. **Hang hutches from the ceiling joists or rafters.** Wire and S hooks work well. Hanging means you don't need hutch legs, which get in the way during manure removal and which can deteriorate over time from contact with manure and urine if not cleaned and repainted regularly. Several hutches hanging together will not move or swing appreciably.

3. **Hang hutches back to back, particularly if you plan to put in an automatic watering system.** Such a configuration helps you keep hutches well away from walls and in the center of the building, with a walkway all around. Automatic watering will be less expensive and more compact if the hutches are centrally located; it will also require less tubing. If the water will be heated or pumped through the system in winter to keep it from freezing, it will require less electricity and it won't have to travel far.

4. **If you plan to have two or more tiers of hutches back to back, leave space between the back walls of the lower tiers.** This allows for clearance of droppings from dropping boards above. One easy way to allow for that — the way I do it — is to use smaller hutches on the bottom tier. In my barn each hutch in the top tier is 30 inches deep, front to back, but the bottom tier hutches are only 18 inches deep. That leaves 2 feet of clearance, which is more than adequate. You could easily use hutches 24 inches deep or even 30 inches deep on the bottom tier if you pulled that tier forward 6 inches, leaving a foot of clearance in either case. The smaller hutches are fine for breeding bucks and growing juniors. The lower tier is usually cooler, too, which is good for bucks.

My Hutch-Hanging Method

I have my own way of hanging hutches in my barn. You could spend a lot of money on a fancy engineered suspension system, and if you are planning a large commercial rabbitry, maybe you should. Or you could try my approach and spend very little. Here's how I do it.

MATERIALS
- 12-gauge wire (available in farm-supply and hardware stores)
- U staples
- 1 × 6 board (cut to length of hutches)
- Corrugated fiberglass panels, 26 inches wide × 8 feet long

For the manure scraper
- 6-inch × 4-inch piece of corrugated fiberglass (cut from leftover piece)
- Wooden handle (can be cut from an old broomstick or dowel)

EQUIPMENT
- Measuring tape
- Screws for scraper handle
- Saw or tin snips
- Screwdriver

Setting Up the Hanging Hutches

1. Suspend the top tier of hutches with the wire.

2. Hang the lower tier from the upper, with 10 to 12 inches between the *bottom* of the top tier and the top of the bottom tier.

3. With the U staples, fasten the 1×6 on end to the suspension wires.

4. With the saw or tin snips, cut corrugated fiberglass panels to fit, allowing for 4 inches of overhang in the rear and in the front to protect the bottom tier from errant urine.

5. Lay the panels on the board and the top of the lower-tier hutches, overlapping as you go by a corrugation or two to prevent leakage to the bottom cages.

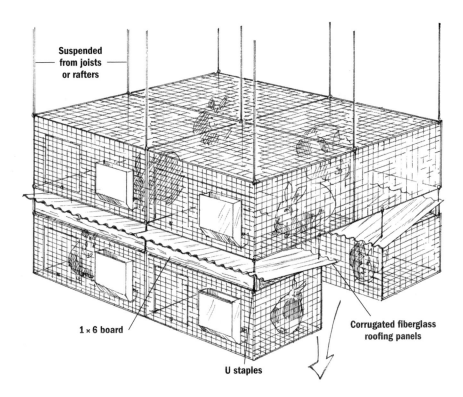

Hanging hutches back to back provides plenty of access. Note also that leaving space between hutches on the lower tier allows clearance for droppings from the boards above.

6. Cut slots in the fiberglass panels to slide around the suspension wires.

To construct a manure scraper that exactly matches the contours of the corrugated dropping boards, cut a 6 × 4-inch piece of fiberglass with the tin snips and screw it to a wooden handle (made from a length of a broomstick or any handy small dowel).

The fiberglass in my barn has lasted 40 years. In fact, it outlasted the first shed it was in, which was metal. It doesn't rot or rust and is easily removed for a good hosing every now and then.

Building a Dream Barn

Never one to lust after fast cars or sleek boats, I've always wanted a barn. It didn't necessarily have to look like a barn, but it did have to act like one. That meant it needed work space and room to store everything a pack rat needs for joyful living. And yes, space for 36 hutches. Not too much to ask, I thought.

But my wife disagreed. She never saves anything for two minutes longer than she needs it. Besides, she couldn't see spending money on it. And if I did build my own barn, as it was beginning to look like I would anyway, it had better look good to her. No shedlike structure was going to mar her landscape — it had to "go with" the house.

While I saved my money, I considered my options. I drove all over looking at other people's barns. I checked out the prefabricated models at the home centers. I examined pictures in books and magazines. I even drew my own pictures. I bought a book of plans. But nothing appeared to fit my needs: a barn that was inexpensive, simple to build, and a good match for the house, which is your basic two-story Colonial with attached two-car garage.

A garage 24 foot square. Hey, that was it. I'd build my garage all over again, but detached. I already owned the blueprints for my new "barn"

The author's dream barn was built from garage plans.

because I'd had the house with garage built a few years earlier. Not only could I use those plans, I'd have the garage standing there in front of me to serve as a full-size working model.

Going this route had plenty in its favor. Specifying the materials would be a cinch, and I could match the roof angle, eave details, door design, and even the siding to the house and garage, which was clad in narrow clapboards with vertical boards and battens on the adjacent porch. Vertical boards and battens: That was it — I'd use wide pine boards and narrow battens. I liked the idea. Even more important, my wife did too.

To hold down the cost, I gave up the notion of a poured concrete slab floor. Besides, I preferred one of sand or gravel. I decided to put in concrete footings and use square 6×6 pressure-treated posts. It was to be, in effect, a pole barn.

A trout-fishing trip in northern Vermont took me past a couple of lumber mills that featured hemlock and spruce framing members and long rows of wide pine siding boards. I realized I could buy them rough-sawn and green and have them delivered to my place for a fraction of the cost of kiln-dried lumber and plywood. Working with green lumber was not my idea of the easiest way to go, but a funny thing happened in my yard while I wrestled with the concrete footings and the posts. Because it was an evening and weekend project, by the time I was ready to saw and nail the lumber, the sun and wind off nearby Lake Champlain had dried every board straight and true.

In fact, it was an evening and weekend project from June through December. By the light of a drop-cord bulb, my freezing fingers pinched the last roofing-shingle nail in a snowstorm on a late December evening.

Except for the wiring and the exterior stain left for the following spring, the job was done. It had taken a mere 7 months. Even better, it had cost very little, less than a fourth of what it might have. Best of all, I was thrilled, and my wife admitted that she actually liked it. The point is that if you want a barn, it doesn't take a lot of money, and the work can be fun if you take your time.

The barn measures 24 × 28 feet, including a 4-foot overhang to cover five cords of firewood. The floor is bank-run gravel. If I wanted to, I could float a concrete slab in there someday. The posts are on 8-foot centers, anchored to the footings below the frost line. (I know they are below the frost line because they have remained square and plumb for over 28 years.) A few 2×10 girts tie the posts together at the bottom. Double 2×10 girts support the ceiling joists and rafters. The joists are 2×8s every 2 feet. Three of them are nailed to 2×6 king posts, and I installed collar beams every 4 feet. These form a great attic area while giving the barn plenty of structural strength. I should add that rough-sawn green lumber comes in its full dimensions: A 2×6 actually does measure 2 inches by 6 inches, unlike the slimmer milled stuff offered by conventional lumber dealers.

I nailed asphalt shingles that match my house and garage to 1×5 hemlock roof decking. If I built it again, I might use plywood on the roof, as the 1×5s required a lot of work. I would also use pressure-treated plywood around the bottom perimeter instead of the double 2×10 skirt boards that I soaked in creosote, but that plywood was not available in

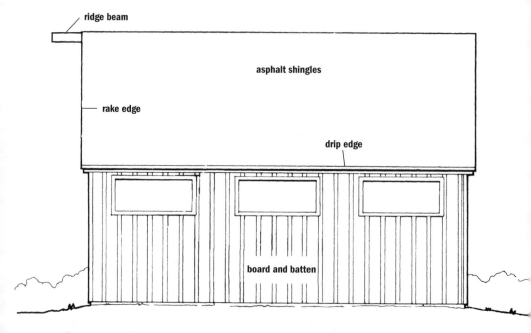

South side of the barn before the greenhouse was added. The ridge beam extends over the hayloft door so rope and pulley can lift hay to loft.

WHERE RABBIT OWNERS' DOLLARS GO

Recently, I conducted a survey of major rabbit-equipment suppliers on behalf of a company interested in entering the rabbit-equipment business. I found out that U.S. and Canadian rabbit raisers spend more than $10 million a year on equipment and supplies, not counting feed. Most of this money was spent in mail-order business or at rabbit shows, but much of it was spent in feed stores, which sell all kinds of equipment and supplies. The most popular items were wire hutches, feeders, watering equipment, and welded wire, in that order.

Twenty-five suppliers to the rabbit industry participated in the survey. I did not include the many online retailers of hutches and equipment, so the $10 million figure is probably quite a bit under what is actually spent on behalf of rabbits.

my area when I built it. I should add, however, that 28 years after I built the barn there was still no sign of rot in the skirt boards.

The rough-sawn, vertical pine boards and battens give it the barn look I sought, and I set the whole thing off with semielliptical doors. This barn could be a garage someday. I spent days figuring out and building the door openings and doors, but it was fun, and my neighbor found some antique iron strap hinges that give them a nice touch.

On the south side I put in three 5-foot-wide windows for lots of light. A few years later, I added an 8-foot greenhouse that spans a third of the south side.

The 36 hutches hang in half of the barn. The other half has plenty of room for all the stuff I really do need, such as two tractors, a rototiller, and assorted other gardening equipment, including carts and tools.

My rabbits have wonderful space, and there's room for plenty of lumber for my next major project. My wife doesn't understand my need to collect and store lumber, and she's been known to say that, except for the rabbits, I built the barn so I'd have a place to keep the boards left over from building it.

Barn Design Plans

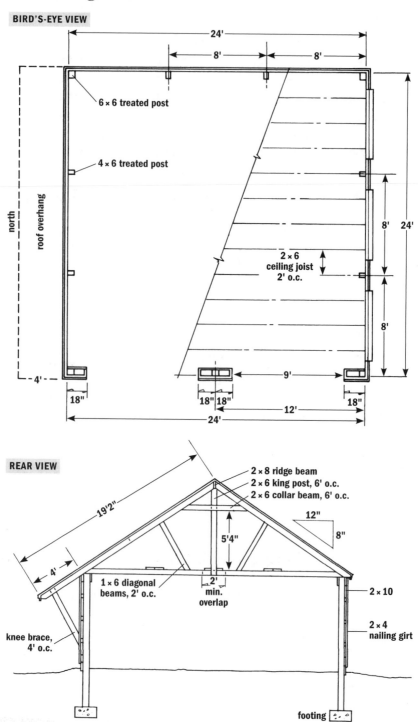

BIRD'S-EYE VIEW

24'

8'   8'

6 × 6 treated post

4 × 6 treated post

north

roof overhang

2 × 6
ceiling joist
2' o.c.

8'   24'

8'

4'

18'

18" 18"   9'   18"

12'

24'

REAR VIEW

19'2"

2 × 8 ridge beam
2 × 6 king post, 6' o.c.
2 × 6 collar beam, 6' o.c.

12"
8"

5'4"

4'

1 × 6 diagonal
beams, 2' o.c.

2'
min.
overlap

2 × 10

knee brace,
4' o.c.

2 × 4
nailing girt

footing

2 × 4 BRACE AND PLYWOOD GUSSET

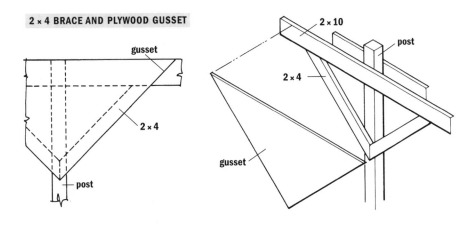

gusset

2 × 4

post

2 × 10

post

2 × 4

gusset

TYPICAL WALL SECTION

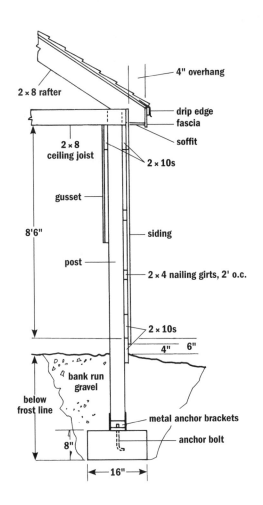

2 × 8 rafter

4" overhang

drip edge

fascia

soffit

2 × 10s

2 × 8 ceiling joist

gusset

8'6"

siding

post

2 × 4 nailing girts, 2' o.c.

2 × 10s

4"   6"

bank run gravel

below frost line

metal anchor brackets

anchor bolt

8"

16"

Rabbit Sheds

You can find many utility or garden sheds on the market today that offer good housing for rabbits as long as they are modified to provide good ventilation. Also, chicken coops no longer in use may be reactivated. Or you can build a shed yourself. A good friend of mine built the shed shown in the illustration on page 56, and it serves him well. (Directions for building it begin on page 68.) He built his shed from perforated-steel angle iron, a very versatile framing material. It requires only nuts and bolts for assembly, and you can find it in most home centers and hardware stores. Not only does it support the shed structure, but it is also sufficient to hold the weight of the hutches and rabbits inside. The angle iron goes together like a child's construction set. Except for sheathing and roofing, no wood was used.

Following the drawing opposite are directions and plans for building a basic 6-foot-high, 10-foot-long, 4-foot-wide, slanted-roof rabbit shed of this kind. You can use these directions as guidelines for building a smaller or a bigger shed, of course, according to your own needs, such as the number of hutches you plan to use. You can also easily add extensions as your rabbitry grows, and I'll fill you in on a few other optional additions. As always, use common sense as your best guide.

SHED MATERIALS CHOICES

Slotted (perforated) angle iron available at home centers and hardware stores was used for the shed shown below, but you could also use:

- **Old steel-shelving framing members.** Watch the classified pages of the local newspaper, which might reveal a warehouse having a going-out-of-business sale.
- **Unperforated-steel angle iron.** This can be obtained used from a junkyard or recycling center. Choose this material only if you have access to a heavy-duty electric drill to make the holes for the bolts.
- **Pressure-treated lumber and plywood.** Nuts-and-bolts construction can be used, as with angle iron.

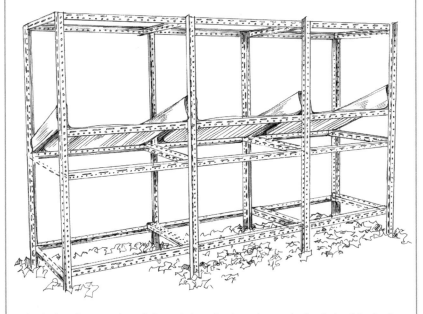

Angle-iron framework and slanted dropping boards are the basis for this shed. Nuts and bolts hold it together.

Slanted-Roof Rabbit Shed Plans

MATERIALS
- 176 feet of angle iron
- Approximately 72 square feet of galvanized sheet metal or corrugated fiberglass roofing panels for dropping boards
- Footings, pressure-treated wooden members, railroad ties, or concrete patio blocks, on which the uprights will rest
- Exterior-grade or pressure-treated plywood, ½ inch or ⅝ inch (you'll need 88 square feet for sides and back and another 40 square feet if you use it for the roof)
- Cedar or asphalt shingles or rolled asphalt roofing to cover 40 square feet
- Paint for finishing

EQUIPMENT
- Measuring tape
- Angle-iron cutter or hacksaw (cutter can be rented from angle-iron supplier)
- Hex nuts and bolts
- Hex wrench
- Nut driver
- Sheet-metal shears

TOP VIEW

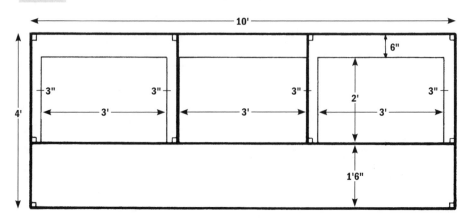

FRONT VIEW

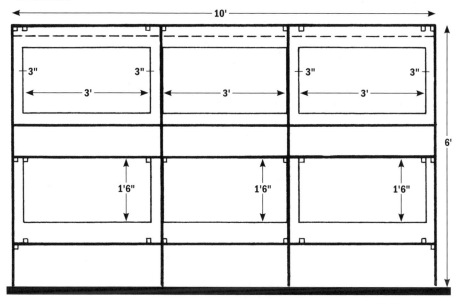

2 × 6 pressure-treated footing (foundation) or railway tie

END VIEW

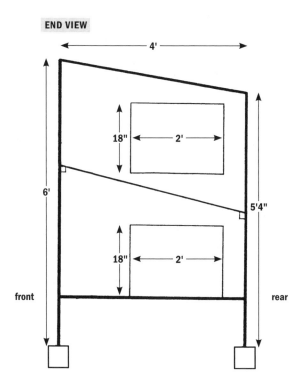

Building a Rabbit Shed

1. Measure the angle iron. Use the slots or holes as guides.

2. Cut all the angle-iron pieces.

3. Assemble the frame, using the hex nuts and bolts, hex wrench, and nut driver (see illustration below). Having another pair of hands is very helpful for the assembly.

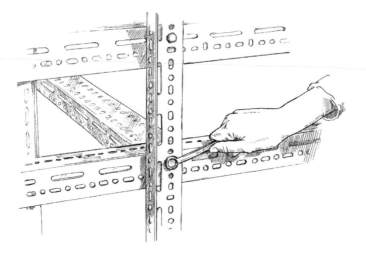

4. Measure, cut, and attach the dropping boards from corrugated fiberglass or galvanized sheet metal. Angle the boards front to back; a good pitch is necessary so droppings will roll to the rear and fall to the ground. Leave enough room in front between the cage bottom and the droppings board so you can use a scraper or hoe occasionally on the droppings board.

5. Bolt the frame into place: Bolt to footings or to pressure-treated wooden members or railroad ties, which you might want to bury underground. If high winds are not a problem, merely setting the frame on bricks or concrete patio blocks can suffice.

6. Bolt the plywood for the exterior directly onto the frame.

7. Make the roof: Use the corrugated fiberglass roofing or plywood; if using plywood, finish with shingles or rolled roofing to assure water-proofing. Allow enough material to provide an overhang in front. The overhang will help keep you dry in the rain while you tend your rabbits; it will also help keep direct sunlight off the animals so they don't become overheated.

8. Finish the exterior with paint to your liking.

Special Features

You can tailor your rabbit shed to suit your purposes, but here are some things to consider:

Cleanout Panel. A plywood cleanout panel in back is handy for cleaning the droppings from the area beneath the cages. Hinge the panel 18 inches off the ground.

Roll-Up Canvas. A canvas or vinyl tarp curtain that can be dropped down in front of the shed will help protect your rabbits from inclement weather. It can be rolled up out of the way the rest of the time. A single unit may be used with a roll-up canvas or tarp in front, or two units may face each other.

Fiberglass Skylight Roof. A white fiberglass skylight roof can be used over facing units, and a more extensive use of fiberglass could be used in the shed's sheathing instead of plywood. White fiberglass admits light but reflects heat. It also lets you view the rabbits in a more natural light than does the green fiberglass. Don't use clear fiberglass, however. It admits more light but also more heat. It's great for greenhouses, but if you use it for hutch sheds, you may have roasted rabbit on the hoof.

Extensions. One nice thing about angle iron is that you can add extensions by simply bolting on more sections to the existing ones. And if you should decide to house your rabbits another way, or simply get rid of them, the whole shed can be taken apart and the angle iron used for shelving in the garage or basement, even for a workbench frame.

Other Rabbit-Shed Options

Countless options are available to you when building a shed. Following are two of my favorites.

Arbor Shed

It never hurts to keep rabbits out of concerned neighbors' constant sight. An arbor, shown below, can be covered with roses or any climbing vine that thrives in your area. The arbor shed requires a roof under the arching top members. You can also plant an evergreen hedge around the perimeter for protection against wind, rain, and snow.

Lath House

Similar to the arbor shed is the lath house, shown on the next page. Such a structure, usually with a conventional peaked roof, is sheathed with latticework. Again, you need some kind of waterproof roof and cold-weather protection. You can find lattice prebuilt of pressure-treated wood and even of plastic, which won't rot and never needs painting.

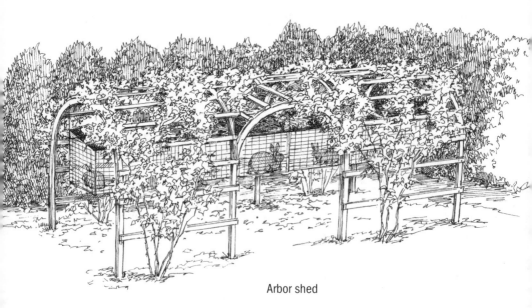

Arbor shed

Lath house

The Economics of Sheds

Depending upon how many hutches you plan to have, a shed may prove cheaper to construct in the long run than individual protection for each hutch. Make certain that the hutches are easily removable from the shed. They can receive an occasional hosing off and disinfecting this way, and the inside of the shed is more accessible for painting. Also, should you decide at a later date to switch to larger or smaller breeds, you will be able to do so easily by hanging larger or smaller hutches inside the shed.

Integrating a rabbit "outbuilding" into your homestead or landscape can make it an asset to your property. For example, I built my barn with roof angles and trim that complimented my house and garage. A building that resembles a tool shed might work well next to a vegetable garden. One that looks like a pool house would blend with your swimming pool.

You can spend very little on rabbit housing if you are enterprising or ambitious. One of my friends told me about a family that bought a house and a wooded lot and a buzz saw. They cut some trees, made some boards, built a shed, and then sold the saw for more than they paid for it. Their shed turned a profit. You might even find a building for free. My local paper sometimes offers a barn or garage or shed free for tearing down or moving. On the other hand, you can make a sizable investment, as in the *Ultimate Rabbitry* example.

Other Equipment and Considerations

What do you need inside your hutch other than rabbits? Three things only: feeders, waterers, and nest boxes. How many nest boxes you need depends on how many rabbits you keep and how many litters are likely to be born around the same time. But of course you don't need nest boxes for

---

RAISER'S EDGE

The Ultimate Rabbitry — Environmentally Controlled

DEBBIE VIGUE built a rabbitry adapted to the cold climate of the far northeastern United States. Here's how she described it:

"Over the past 16 years we've had a few different setups," she said. "We started with four holes [rabbit jargon for cages] enclosed on three sides with an overhang on the open front. That was because my husband, Gary, reminded me I was to have only two, *only two* rabbits.

"Next we progressed to a 15-hole hanging-cages building with half walls that could be taken out in the summer. Then we overflowed to 21 stacking cages in the horse barn (oops, sorry, Gary). After many winters of frozen water dishes, we built a brand-new rabbitry. I love it. We had a concrete slab put down and a 24 × 32-foot building erected. It has nine windows; a metal, insulated nine-light entry door; and a solid, 4-foot-wide door in the back for a wheelbarrow to fit through. There are two louver vents in the center, front and back, in which are fans.

"The front fan pulls in air and the back fan sucks it out. An oil furnace in the center keeps the temperature from falling below 47 degrees in the winter. Four heating vents at the bottom blow warm air below the cages. The ceiling is 8 feet high. There is 16 inches of insulation in the attic and 4 inches in the walls, which keeps the heating bill reasonable and the summer heat out, with help from the concrete floor.

each hutch, as they can be rotated when and where needed. The young stay in the box only 3 to 4 weeks.

You also need to think about ventilation and about what demands, beyond basic shelter from heat and cold, your particular climate will make on you and your rabbits.

Feeders

I've known rabbit raisers who use a 1-pound coffee can for a feeder. They say they can't afford to use the kind of galvanized metal self-feeder that I use — which costs only a few dollars. Those who claim they can't afford to buy this feeder should take a look under their hutches. There they'll

---

"The front countertop has a stainless steel sink with running cold water, a grooming rug, and a four-drawer supply cabinet. Shelving underneath stores extra food and water crocks, tattoo supplies, carrier feed cups, and bottles. A fly spray mister sits on one shelf and is used only occasionally, as few flies get inside. Sometimes we use an air freshener in the mister.

"The back corner has an enclosed shavings bin and a pop-out opening that allows us to shovel the shavings into the bin from the outside. Two rows of fluorescent lights are centered over the two cage rows. Another strip is over the front counter area. One row of three light bulbs goes down the center. We leave them on during the day. We painted the walls light blue to lighten the inside. We stained the window and ceiling trim. An indoor/outdoor thermometer with a digital readout not only shows the temperatures but records the highs and lows indoors and out. A wireless intercom system to the house keeps us in touch.

"Although stacked cages are labor-intensive, I prefer them. We have 60 holes that are seldom completely full. It takes the two of us less than two hours a week to clean and sweep up. The building has worked out well for us. It is nice to go out to the rabbitry on a cold Maine winter day and work in a sweatshirt. It is also comfortable in the summer. I know I spend more time with them now and that the bunnies are more comfortable, too."

find enough wasted feed to pay for a feeder — maybe even one made of sterling silver.

The rabbits tip coffee cans over or scratch out the feed. They get two to three times as much feed as they need, and yet they don't get enough to eat and are underweight. When droppings fall through the floor of the wire hutch, so do rabbit-feed pellets.

The self-feeder, which attaches to the outside of the hutch, has a lip on the rim of the trough to prevent the rabbits from scratching feed out, and of course, it can't be tipped over. It doesn't take up precious space on the hutch floor, and it can be loaded from the outside of the hutch, without opening the door. Rab-

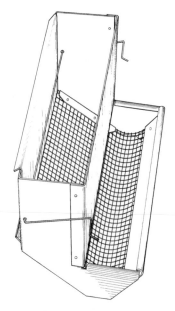

A screened-bottom self-feeder will pay for itself in feed savings.

bits cannot easily foul it because they can't sit in it, so the feed stays clean. As a bonus, it holds enough feed for a couple of days in case you have to be away.

It's true that there are some earthenware crocks to be found that won't tip over and that have a lip to prevent rabbits from scratching out the feed. But crocks do take up floor space and must be constantly cleaned; some feed is always soiled and wasted. So a metal self-feeder, especially one with a screened bottom that lets pellet dust sift through, gets my vote.

Waterers and Watering Systems

Rabbits need clean, fresh water at all times to grow well. For a number of years, I used earthenware crocks to water my rabbits. I've never used tin cans, because, just as with cans for feed, they tip over frequently, and the rabbits go thirsty. Of course, tin also rusts. An earthenware crock isn't ideal; it needs a lot of washing and daily rinsing, but it comes a lot closer than tin cans.

Yankee Ingenuity in Texas?

P HIL ANGELL, an ingenious Dallas rabbit raiser, is always seeking to do things more efficiently. Not long ago he devised a combination hay/pellet feeder, which is now manufactured by Bass Equipment Company (Monett, Missouri).

When he fed hay with the standard hay rack, Phil saw much of the leafy and nutritious part of the alfalfa hay fall through the wire flooring of his hutches, wasted on the ground below, while his rabbits munched mainly on the stems. So he added some metal side extensions to a hopper pellet feeder. The space provided above held the hay. What didn't get eaten fell into the feed trough below to be eaten later.

Years earlier I noticed the same problem and simply built my hayracks inside the hutches and over the feeders. But Phil's idea is much better because it saves more hay and eliminates the hay rack altogether. After he had made a few of these combination hay/pellet feeders, he sent me one. I liked it so much I sent it to the Bass Equipment Company, which put it into production and made it commercially available. The Bass version features a larger-than-normal feed trough to permit full access to both pellets and hay. Your feed store may have the hay/pellet combination feeder in stock or can get it for you. I use them in my rabbitry now. You have to cut the hay into short lengths, but that's a good idea anyway, because rabbits love to pull long stems from a hayrack, nibble on some, and let the rest fall through the wire floor.

An automatic watering system pipes water from a city water system or a well into a 1- to 5-gallon breaker tank that reduces the pressure and then, using the force of gravity, down through tubing to the valves in the hutches. A semiautomatic system features a holding tank, which might be a large jug or even a big pail, and tubing leading to the hutches. You have to fill the holding tank yourself, perhaps with a hose, but you don't have to fill individual jugs or bottles or crocks (and you don't have to rinse and wash them). Both kinds of systems use the automatic watering valve.

Using the Automatic Watering Valve

When I discovered the automatic watering valve, I packed my crocks away. This little valve costs less than a crock. It was designed to be inserted into pipe or flexible plastic tubing that carries water from a tank to the hutches. That's the way I use these valves now, in a manner I will soon describe, but earlier I used them without the tubing, inserted into plastic milk jugs or soft drink bottles.

With this method, the jug or bottle hangs outside the hutch with the valve protruding inside. The top is left off to allow air in, required to make the valve function properly. The rabbit simply bites on the end of the valve, which has a little stem inside it that swings aside to let the water trickle into the animal's mouth. When it stops biting, the valve closes and the water stops.

The Valve-Jug Method. To insert the valve into a plastic half-gallon bottle or jug, cut a small hole with a sharp knife, and press the valve into

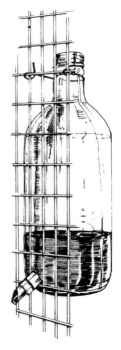

place. Seal the opening with epoxy cement; that keeps water from leaking and the valve from falling out. It will provide your rabbits plenty of drinking water that stays fresh and clean. Using the gallon or half-gallon bottles or jugs, you need to replenish the water only every other day instead of every day, as you would with a crock. If you should be away for a day, nobody has to water your rabbits for you. Ready-made plastic-bottle waterers with valves attached are also available, although these will cost a couple of dollars or so more than the valve alone.

The Valve-Tubing Method. The little valve does a fine job with a jug, but it works even better when attached to tubing. Flexible plastic tubing costs very little and can be fitted easily with these valves. You can set up this system yourself with ease. I had no

Valve-jug method

R A I S E R ' S   E D G E

More Texan Ingenuity

PHIL ANGELL of Dallas, Texas, is always thinking. He's come up with what he calls a "simple water filter" for a homemade automatic watering system.

"All you have to do," advises Phil, "is go to the local pet shop and buy an air stone for about a dollar." This device is simply a man-made porous cylindrical unit about 1 inch long and ½ inch in diameter, which has a nipple that fits an ⅛-inch air line and keeps sediment from entering that line. It's supposed to aerate the water in a fish tank when attached to an air pump. To create the water filter, simply get an inch or so of ⅛-inch tubing, push it onto the air stone nipple, and then slide it into your watering system line, which goes in the holding tank. "Works great," he said.

experience with this tubing when I put together a completely automatic watering system for my rabbits, using only a pair of scissors and a tape measure.

To put together a tubing system, run the tubing from a 1- to 5-gallon breaker tank or bucket to the hutches in one continuous line from the first to the last hutch. At each hutch cut the tubing, and insert a tee connector and another piece of tubing with a valve pushed onto the end. Use a bracket to attach the valve to the hutch, and use standoff clips to hold the tubing away from the cage wire so that the rabbits can't puncture it with their teeth. Insert a drain valve at the end of the tubing.

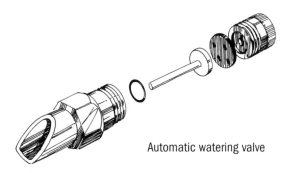

Automatic watering valve

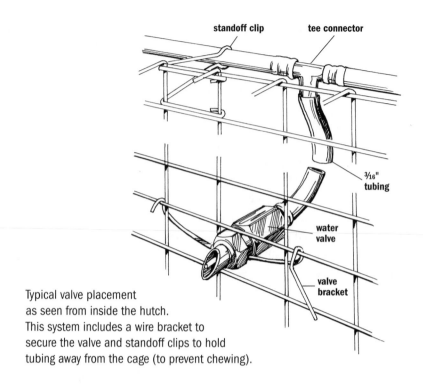

Typical valve placement
as seen from inside the hutch.
This system includes a wire bracket to
secure the valve and standoff clips to hold
tubing away from the cage (to prevent chewing).

It is important to get the tubing, the valves, and the fittings from the same supplier so that all the components are compatible. Use black-plastic tubing, not transparent. Algae can grow in transparent tubing even in artificial light.

Winter Watering

Below-freezing temperatures complicate watering for obvious reasons. Because your rabbits require water daily, err on the side of caution, and follow these watering recommendations.

Crocks and Jugs

If you do use a crock in winter, look for one that has a smaller interior diameter at the bottom than at the top. Otherwise, if the water freezes in winter, you will be left with a cracked or broken crock, because water expands as it becomes ice. Fill them only halfway in the morning, let the rabbits drink, and either dump out the water or let it stand and freeze. Then at night, fill the crocks again and hope that the rabbits get a good drink before they start licking ice. The next day, take the crock inside, or put it into a bucket of hot water to remove the ice. That is a nuisance, but it must be done.

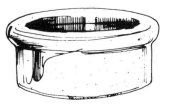

Watering crock

Clearly, it's nice to have two crocks for every hutch, so one can thaw while the other is in use. But that can be expensive. One-quart (1 L) metal watering pans with tabs to hold them to wire-cage walls are a good alternative, and you can smack the ice out of these, too. If you use jugs with valves, you can provide two for each hutch, alternately using one and thawing the other.

Semiautomatic and Automatic Systems

With a semiautomatic or an automatic watering system, you can keep the water from freezing by using an immersion (bucket) heater in the tank or a small electric pump to keep water moving through the system like a tumbling mountain stream, or both. These heaters and pumps are available at farm-supply stores and come with instructions.

You can of course simply drain the tubing on cold nights, particularly if you live where freezing temperatures are rare. You could use crocks or jugs at those times. With plastic tubing there is no need to fear breakage when the water freezes within, because the plastic expands to accommodate the ice. Of course, if your rabbits are indoors, you could heat the building. One way or another, you can and must keep your rabbits watered.

WATER REQUIREMENTS

Most rabbits drink a quart (or liter) or so of water per day. In hot weather they drink more and in cold weather, less.

Nest Boxes

The one other piece of basic equipment you really need is a nest box for the doe and her litter. A box 18 inches long, 10 inches wide, and 10 inches high is big enough for all but the giant breeds, which need a box about 24 inches long and 15 inches wide. Make certain that the box will fit through the hutch door. You can construct it of light plywood or ¾-inch lumber, or you

The wire nest box has a corrugated cardboard liner. A metal flange on the top edges holds the liner in place and can be removed when a new liner is inserted for the next litter.

can purchase a metal box. Perhaps you will be able to find an apple or wine box that you can adapt to do the job. Don't build or buy a covered box. These are damp, which can be fatal to your rabbits. I don't care how cold it gets where you are; an open box is better than a closed one.

A popular style of box is one made of welded wire, such as is used on the floor of rabbit hutches, and lined with cardboard in cold weather. Don't use a cardboard box alone, however, as it may turn over, destroying the nest and the babies along with it. Rabbits will also chew it to pieces in short order. Always have a new, clean cardboard liner for the welded wire box. In warm weather use cardboard on the bottom only, so air can circulate through the wire-mesh sides. I have found that the wire nest box, like the wire cage, is the most satisfactory, and I now use it exclusively. I bought some and also built some myself. I make my own cardboard liners from discarded corrugated boxes.

Ventilation

If there is a last word on housing, it is *ventilation*. Good ventilation is perhaps the most essential and most overlooked aspect of hutch, shed, and barn construction. The wire hutch provides wonderful ventilation, but if it is in a building of any type, that building must have great ventilation, too. My barn in northern Vermont is unheated, but I leave the windows open all winter, except during occasional bouts of blowing snow. If you keep your rabbits out of the wind, the rain, the snow, and the blazing sun, they will be fine, provided that they have adequate ventilation. You can provide this naturally, or you can install fans.

My barn windows are high on the south side and there are air outlets low on the north side. Prevailing breezes are sufficient for good ventilation. Prevailing breezes may work for you, too. If not, install a fan and exhaust the air near ground level after bringing it in from above. Don't pull air up from the ground or floor, through droppings and urine with its attendant ammonia. Push those fumes down and out!

Ideal Temperature

Ideal temperature in a heated, air-conditioned, and insulated controlled environment would be 50°F (10°C). But remember that rabbits can take just about any amount of cold, because their fur keeps them warm. When temperatures get above 90°F (32°C), some rabbit raisers fill plastic jugs with water, freeze them, and place them in hutches to help cool their animals. Good insulation and ventilation are vital in hot weather. Fans are sometimes useful. Misting fans attached to garden hoses are quite effective in extremely hot temperatures.

A NOTE FOR NAYSAYERS

Nobody, not even the most avid animal-rights fanatics, needs to worry about depriving caged rabbits of their liberty. Those born in confinement don't know what liberty is.

Also, you will note I advise housing rabbits only in wire hutches. If you house them this way, you and the rabbits will do just fine. But if you let a rabbit out, you are asking for trouble. A rabbit that runs around a house will gnaw furniture and electrical cords, though it will gnaw the cords only once. Then you will need new furniture *and* a new rabbit. You might even need a new house if a rabbit gnaws through a computer cord and sets the place on fire, a sad situation that occurred not long ago. This goes for pet rabbits too. The wire hutch is ideal. Keep them there, except perhaps to hold and pet them.

5

Feeding Rabbits Right

EVERYBODY KNOWS WHAT RABBITS EAT, RIGHT? Cabbage and carrots and plenty of 'em, right? *Wrong!*

There are plenty of things that rabbits *will* eat, including the salad items mentioned above, but to be successful with rabbits, or any other livestock, you must feed them what they *should* eat.

Just as nutritionists have determined what humans should eat, livestock nutritionists have concluded that rabbits need a balanced diet to grow and reproduce. Because you will be trying to put your rabbits in the best of health for optimum growth and reproduction, you will want to feed them the optimum quality rations.

Pellets Are the Ideal Feed

The pellets now available in sacks got there only on the basis of a complete understanding of what the rabbit requires. The pellets contain *everything* the rabbit needs:

- Alfalfa hay for high-quality roughage
- Special sources of protein, including some of animal origin
- Phosphorus, calcium, essential trace minerals
- Necessary vitamins

Protein levels of these pellets range from 12 to 20 percent. A 16-percent protein level is sufficient for milk production in a doe bringing up a litter and for a growing rabbit. Higher protein levels really are not necessary. My preference is for the 16-percent feeds, which have high-fiber percentages to help prevent digestive problems.

The protein content and the ingredients of different brands of pellets may vary slightly; in fact, the contents of the same brand may vary from feed mill to feed mill. But all the varieties are good, and one thing is certain: if all you ever give your rabbits are pellets to eat and clean water to drink, they, and you, will do fine.

Research Continues

The major feed companies conduct ongoing experiments and have advanced the feeding of rabbits to an extremely high degree. Mortality of young rabbits has been reduced tremendously since the days before pellets were available. The rabbit growth rate has soared.

Yet despite all of the ironclad evidence that pellets are the best feed for rabbits, there still are those who insist on feeding rabbits something else. Of course, rabbits can exist and even thrive on other feeds, but the odds are that they won't. You might think you could feed rabbits for less money than pellets cost, but it is not very likely if you consider death losses, longer times to reach market weight, poor-quality carcasses, and fewer and smaller litters. In addition, quite often nonpelleted feed is wasted because it spoils easily or because rabbits scratch it out of feeders (sometimes, as in the case of mixed grains, they prefer what's down under to what's on top).

A covered metal garbage can keeps pellets clean and dry and keeps mice out; a scoop is handy for feeding.

IT USED TO BE A TOUGH JOB

Many years ago it was difficult to gather all the feed grains, roots, and roughage that rabbits need. But now feed manufacturers, backed by degreed professional research nutritionists, have assembled all the ingredients for rabbit raisers. And they have sacked them up for you in the form of the bite-size rabbit pellet.

Feeding Simplified with Pellets

You can make feeding rabbits a complicated affair with unpredictable results, or you can simplify it so that you can count on your rabbits performing the way you hope they will. If you feed pellets, it is an easy matter to calculate how much to feed and when. If you feed anything else, judging the amount is a difficult task because nonpelleted feed is not easy to measure.

Feed Supplements

While I strongly urge you to feed pellets, I'm not averse to supplementing them with other items, but these must be used judiciously. You're better off not trying them until you get used to your rabbits. Supplemental foods (covered in chapter 8) will complicate your feeding schedule. Nevertheless, there are times when such foods can help you, depending upon your objectives and what is available to you.

You can put together your own feed based on the guidelines in the above chart, using three sources: dry roughages, greens (including roots and tubers), and concentrates.

Dry Roughages and Greens

Dry roughages include alfalfa hay, clover, lespedeza, oat hay, peanut hay, soybean, timothy, and vetch. Greens that provide sources of rabbit feed are carrot, rutabaga, sweet potato, and turnip greens, along with lettuce. Wash all greens thoroughly.

WARNING!

Don't *ever* feed greens to young rabbits. Greens can give the youngsters diarrhea and can kill them.

Concentrates

Among the concentrates are barley, dried beet pulp, bread, brewer's yeast, buckwheat, corn, cottonseed and linseed meals, milk, oats, peanut meal, sorghum meal, soybean meal, and wheat. If you like, you can mix up rations from these items, calculating the percentage of protein, roughage, and fat they will deliver to your rabbits. Some of the grains will have to be ground, cracked, or rolled. You should feed concentrate rations in a feeder with separate compartments or in separate dishes, because the rabbits are likely to scratch them out, preferring some to others, and waste them if you don't. Then again, you have to watch how much each rabbit eats. Some will eat more of one thing than they should and less of another. Obviously, this is complicated. Pellets deliver all required nutrition in one simple package, which is why I prefer them.

What about Hay?

Alfalfa hay is my first choice for a hay supplement. I know the rabbits enjoy it, even though they already get a lot of alfalfa in the pellets. If they don't clean up all their pellets, they don't get the hay. Almost any other fine-stemmed, leafy, well-cured hay, such as those listed under Dry Roughages and Greens, that is free from mildew or mold also will do nicely. Cut it into short lengths to reduce waste.

General Feeding Guidelines

Young, growing, unweaned, and just-weaned rabbits should be fed all the rabbit pellets they will eat in a day. You will soon find out how much to feed. A lot depends upon the size and age of your rabbits. If feed is left over

WHAT RABBITS NEED

Which Ones?	Daily Nutritional Requirements
Dry does, herd bucks, and developing young	12–15 percent protein 2–3.5 percent fat 20–27 percent fiber
Pregnant does and does with litters	16 percent protein 3–5.5 percent fat 15–20 percent fiber

Mr. Reynolds' Special Rabbit-Feed Recipe

A FORMER PRESIDENT of the American Rabbit Breeders Association, Oren Reynolds, shared with members a special mixture that he felt does a good job of supplementing rabbit pellets. If you are interested in making the effort to feed rabbits the way he did, here's how it goes:

> 6 quarts oats
> 1 quart wheat
> 1 quart sunflower seed
> 1 quart barley (whole if available, otherwise crimped)
> 1 quart kaffir corn (when available)
> 1 quart Terramycin (Pfizer, New York City) crumbles, an antibiotic

1. Mix oats, wheat, sunflower seed, barley, and kaffir corn.
2. Add Terramycin crumbles to feed mixture once a week.
3. Feed one part of this mixture to three parts of pellets daily.

"Since I'm retired and have time," Oren said, "I feed the pellets at night and the grain in the morning. They can be fed together, but some animals like one ingredient better than others and will scratch out the other feed to get to it." It should be pointed out that Mr. Reynolds began his feeding procedures before rabbit pellets were commercially available and perhaps even before the pelleting machinery was invented, for all I know. He raised rabbits almost until he passed away at the age of 101.

the following day, give less. If a rabbit greets you at feeding time by diving headlong into its feed, give more. Adult rabbits of the medium-weight group consume about 5 ounces (142 g) of pellets a day; again, give more or less depending upon the individual and the ambient temperature. Calories are units of heat: in a cold environment, feed more; if it's warm, feed less.

Females with litters should have pellets available to them at all times. But watch out for overly fat does, especially those five to six months old and soon to be mated. Lazy bucks are probably overfed. In fact, most rabbits are probably overfed. Except for does with litters, rabbits do not need feed in front of them constantly. My adult bucks and my does without litters usually clean up all their pellets within an hour, and that's all they get until the next day. Of course, they almost always have hay to munch on.

Watch Carefully at Feeding Time

A good raiser watches stock closely at feeding time. While they are eating, run your hand over each rabbit. A rabbit that's a bit bony should get more feed. If a rabbit hasn't cleaned up its pellets, something is wrong.

To determine the cause of the appetite loss, first check the water supply. Is the crock or jug empty? Is the valve plugged? Rabbits don't eat when they are thirsty. They must have plenty of water. Most of the time a rabbit that isn't eating isn't drinking. Note the droppings under the hutch — they should be large, round, and well formed. Watch out for diarrhea. If a rabbit is off feed, simply not eating well but apparently healthy, try tempting it with a tidbit of dry bread, a leaf of lettuce, or a spoonful of oats. That will usually get it back to the feeder.

Overall, remember this: Your rabbits should be happy to see you at feeding time. When they hear you open the metal can with the feed, they should come bounding to the front of their cages, waiting impatiently at the feeder. They should be hungry and ready to dive into the pellets.

If you run your hand over each rabbit at feeding time, you will know if you should increase the ration or not.

If not, and there are no problems as described above, you may be feeding them too much, so cut down a bit today and see how they greet you tomorrow. On the other hand, if they are ready to tear off the cage door to get at that metal can of pellets, give them a little more. There are no exact rules. It's something that you will learn over time.

When to Feed

If there is one best time of day to feed rabbits, it's in the evening. They are more active at night and will eat more readily when day is done. But if morning is more convenient, your rabbits will adjust to your schedule. Rabbits do, however, require regularity. If evening is your choice, then feed them regularly every evening. Don't make your rabbits wait to be fed, and don't feed them early just because you happened to get home a little sooner. Many livestock raisers like to feed their animals before they sit down to their own evening meal.

If you use water crocks or metal water pans, rinse them daily and disinfect weekly. This routine, in an all-wire hutch that keeps itself clean, will go a long way toward keeping your rabbits healthy.

In chapter 8 I'll discuss some feeds that you can produce in your own garden to supplement pellet feeding. Gardens provide lots of good opportunities to stretch your pellets, particularly with dry does and bucks you are trying to maintain until you can make room in your freezer.

My Preferred Lifetime Feeding Regimen

Rabbit pellets supply all the dietary needs of your rabbits, so why let them fill up on feed that doesn't do the complete job? You wouldn't munch on candy all day and expect to be well nourished. So think of greens, carrots, leftover dry bread, and other items that might be used sporadically as you think of candy — tasty morsels given sparingly as treats.

Here's how I feed my rabbits and how I think you should feed yours if you want to achieve success with them.

The Doe: Before and During Pregnancy

Until a full-grown, adult doe is pregnant, she gets only enough feed to keep her slim and trim. I don't want her fat for two reasons: First, if she isn't pregnant, extra internal fat can choke off the fallopian tubes, stop the descent of her eggs, and thus prevent conception. Second, if she is expecting, extra fat can make delivery difficult and could even kill her.

My does, the Tans and Florida Whites, get 2 or 3 ounces (57 or 85 g) of pellets a day until I know they are pregnant. Then they get another ounce if they will clean it up, and also get about a tablespoonful of oats. In cold weather they get a little more of both, because they need the heat that the calories provide. (Calories are units of heat.) They need more calories in my unheated barn in the winter than they do in the summer.

I don't consider my does pregnant until I palpate them, or feel for the embryos, a procedure I'll describe in chapter 6. Of course, they may well be pregnant, but until I palpate does, I consider them quite unpregnant. And that means I limit the amount of pellets they get.

On the Day of Delivery

When this doe delivers her litter, or kindles, as rabbit raisers say, she gets the pellets, oats, and hay, but she also qualifies for the special treats I have mentioned previously. I offer her a slice of apple, or a carrot, or a few leaves of lettuce, the tops of the celery, or a few potato peelings but not cabbage leaves, which I loathe because of the gassy aspect. This doe is, of course, very thirsty, so in addition to the water that all my rabbits have access to continuously, I provide morsels of greens or roots. I may continue to give her these treats daily until the litter comes hopping out of the nest box. Then the treats cease (although I've been known to feed a bit of them out of my hand to the doe) because I don't want the youngsters to get hold of any greens. Remember young Peter Rabbit? He returned home to a sickbed after his escape from Mr. MacGregor's garden. Outside of an upset stomach, all you can possibly do by feeding greens to a three-week-old rabbit is kill it.

The Doe and Her Litter

I provide the doe and litter with all the pellets they want. That means I keep the self-feeder loaded to the point where it never runs out. They have all they want because the young should grow at a steady pace, and the doe needs to keep up her milk supply while the litter is nursing.

In addition to hay as a supplement, I sometimes feed whole oats — the best I can get. These usually are called racehorse oats and are sold by the 50-pound bag. I feed these in a separate feeder or on one side of my wide self-feeders. When the litter leaves the nest box, at about 16 days, that is the time to add oats. Lots of successful breeders add oats to keep the young growing well, particularly if it's an unusually large litter or if they plan to rebreed the doe earlier than normal (see chapter 6).

What about the Weanlings?

When I wean the litter, which I do at 6 to 10 weeks (see chapter 6), I give them all the rabbit pellets they will clean up in one day. I don't let any feed hang around in the dish and get stale and moldy, by any means. But I make sure that these young, growing animals get all they can put away. If they do not seem to be gaining as well as they should, I give some oats, and of course, I always give hay. I particularly like to give hay to the youngsters, and I believe that you really can't overdo the fiber for the young. The more fiber you give the weanlings, the less chance they have of getting diarrhea, which can be a big problem. In fact, my pellets are no higher than 16 percent protein and are high in fiber. Once again, I never give any greens to rabbits less than 4 or 5 months of age under any circumstances. Greens won't do a bit of good and can kill them.

The Buck

I give my bucks 2 or 3 ounces (57 or 85 g) of rabbit pellets per day, again depending somewhat on the temperature. These are Tans and Florida Whites, remember, which are small rabbits. Some general guidelines for

WHAT ABOUT SALT?

The question of how and when to feed salt often leads to confusion. Rabbit pellets already contain salt, so I haven't fed extra salt in years. If you use a salt spool, it will drip moisture and rust your hutch floor. If you really must add salt because you think the rabbits aren't getting enough, sprinkle it right from a shaker onto the pellets or other feed.

all sizes of rabbits follow, but for now the main thing to remember about feeding bucks is to keep them slim and trim. I don't like to satiate the appetite of the buck, because I want him to retain his appetite for romance. A breeder buck that sits around in the corner like Ferdinand the Bull with no sexual appetite isn't going to do me any good. If I fill him up so he's fat and lazy, I don't think he'll chase females. So I keep him in loving trim.

The Juniors

Junior does, those developing does older than weanlings — oh, 4, 5, or 6 months old, just too young to breed — get a measured amount of feed that will keep them on the slim side as well. If they are too fat, they won't conceive, or if they do, they might have trouble on the day of delivery. So their pellets are limited to 2 or 3 ounces (57 or 85 g) a day, depending on their age and size and the ambient temperature. Remember that calories are units of heat. If it's hot, give less; if it's cold, give more.

I don't worry too much about junior bucks. For one thing, they don't have to kindle a litter. So I let them eat about all they will clean up in a day. But I certainly don't let any feed hang around from day to day.

The Show Rabbits

Conditioning a rabbit for a show is a whole other ball game that I'll cover in chapter 10. There you'll discover that I violate some of the rules I've laid down in this chapter: I feed my show rabbits all kinds of things in addition to rabbit pellets, the kinds of things that the top winners for years have fed their champions. In that chapter I'll offer tips on how to put a keen edge on show rabbits — an edge that can dull quickly, but one that when sharp, will cut the competition down.

What Else Should You Know?

There's nothing as important as good feed and good feeding practice. You can't go wrong with pellets, and the choice between them and a home-mixed feed is a no-brainer as far as I'm concerned. If you are a beginner, I can't urge you too strongly to stick to pellets, especially the lower-protein varieties. As you gain experience you may try other feeds, but make any changes gradually, to avoid digestive problems in your animals.

6

Breeding and Producing Rabbits

IF YOU STARTED RIGHT, you have chosen a breed, obtained foundation stock, and housed your rabbits in all-wire hutches, where they are happily munching pellets. You have done a lot for them so far. But what have they done *for you* lately?

Probably what they have been doing, if you have been taking good care of them, is growing to breeding age. Now they are ready to start producing for you.

Mating Your First Pair

To get into production, you must mate that first pair. For the small breeds your does and bucks should be at least 5 months of age. The medium breeds should be 6 months, and the giants must be at least 8 months of age. For my Tans, a small breed, I always say start breeding at 5 months or 5 pounds (2.3 kg), whichever comes first.

Look 'Em Over

Let's assume you have decided which buck and which doe are to be mated, based upon objectives of your breeding program, which we will soon cover. For now let's agree that you have chosen two specimens from which you wish to produce a litter. Here's how you check them out to make sure they are the right ones to use:

First, look at the doe. She should be at just the right weight, not too fat. Her fur condition should be excellent — no shedding and plenty of sheen. Next, check the vulva. For best results it should be a reddish purple color, not a pale pink. If it has the deep color, the doe probably is ready to breed, provided she is in perfect health.

Now look at the buck. His fur, too, should be in good condition; his coat should shine, indicating his health is good. His eyes should be bright, another indicator. Whether he is fat or thin is not as important as it is for the doe, but if he is too fat, he may be too lazy to service the doe.

Next, check his testicles. If they are completely descended into the scrotum and the scrotum is full and large, he is a good buck to use. If he has only one testicle, or if one or both are withdrawn into the groin or have a withered look, he may be sterile. Even if he isn't, any litter he produces may be small. This condition may be only temporary: one testicle may not yet have descended into the scrotum. Or the withered, wrinkled look may be seasonal. It happens particularly to older bucks in late summer and autumn. But it is extremely important to check this out, to make sure you get the litter you want. If you want reproduction, make certain the reproductive apparatus is in perfect working order — at least as far as you can see.

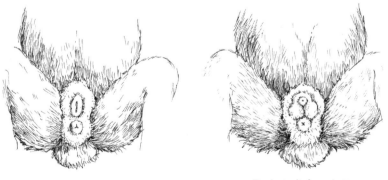

Doe genital anatomy   Buck genital anatomy

Take the Doe to the Buck

If both prospective parents look just right to you, take the doe to the buck's hutch and put her in. Never take the buck to the doe, or they may fight, as the doe can be very defensive about her quarters. And never simply leave the pair alone together. Put the doe in with the buck, and stick around to observe the mating.

Don't blink, or you may miss it. Rabbits are rabbits. They mate like rabbits. If everything goes right, and it usually does, it will be all over before you close the hutch door. The buck will have mounted the doe. The doe will have raised her hindquarters. The buck will have serviced the doe and fallen over backward or on his side.

Now remove the doe. Turn her over and check the vulva to make sure the semen has been deposited there. There's nothing like seeing for believing. Leave nothing to chance.

What about a Cycle?

Is mating really just this simple? Usually, but not always. It has been reported that the female rabbit has no cycle, that she is fertile 365 days of the year. It has also been reported that the female rabbit is fertile for 12 days, followed by 2 to 4 days during which she will not conceive, followed by another dozen days when she will. Either way, cycle or not, a doe will conceive most of the time. If a doe rabbit has a cycle, you really aren't going to know when she is fertile or infertile anyway. So I say, forget it.

It is known that the doe rabbit's eggs descend for fertilization upon sexual stimulation; in other words, after service. This process takes 8 to 10 hours, at which time the eggs meet the sperm and conception takes place. That assumes the sperm is there. If the doe urinates in the meantime, the semen may wash away.

Put Her Back In with the Buck

I put the doe back in the buck's hutch 8 to 10 hours after the first successful service. If the eggs are descending, now's the time for conception. The buck, of course, is ready, and another service may be just what is needed, particularly if the semen from the first service is no longer in there. That is something you won't really know unless you stand around and watch for 8 hours. Like me, you probably have something else to do.

Suppose Nothing Happens?

Okay, you have put the doe in with the buck for the first time, but no mating occurs. Either he sits in the corner like Ferdinand the Bull and pays her no mind, or he chases her like crazy and she resists these advances with unparalleled virtue. You have a problem but not an insurmountable one, to be sure.

If the buck is not interested, try another if you have one. The second is almost certain to service the doe. Finding two unwilling buck rabbits in a row is almost unheard of. But what if you have only one buck to mate? Try this: Pick the buck up and put him on the doe's back. He probably will get the idea. If that doesn't work, leave the doe in his hutch, but take him out. Put him in hers, and leave them there overnight. He will pick up her scent and by the next day doubtless will be more interested. At that time take her to her own hutch and put her in; the buck will probably service her. She also should be more interested, having acquired his scent from his hutch.

If that doesn't work, the buck probably is too fat and lazy. You can start to slim him down by reducing his feed intake, but in the meantime attempt the mating before the buck's next feeding time. Try the morning or the evening before he dives into his pellets.

It is more likely by far that the doe will show no interest. For every buck that isn't ready, one hundred does are not. She may hunch down in a corner of the buck's hutch. She may flatten right down on the floor. She may climb one wall. She may resist every advance instead of raising her hindquarters and lifting her tail to accept the buck.

There are a number of ways to get this doe bred. In my many years of rabbit raising, I still haven't met a doe I can't get bred! Surely, there are those that nature has not completely equipped for reproduction. Certainly, it can happen. But it hasn't happened to my rabbits yet. I hate to repeat it so often, but rabbits are rabbits are rabbits. They will outproduce almost anything.

Here's something else to keep in mind. Sometimes when you put the doe in with the buck, she will run from him and act as though she has no interest in mating whatsoever. But watch her tail. If she twitches her tail, you can be sure you are seeing a coy doe playing hard to get. A doe whose tail is twitching is a doe that's itching for romance. She may run for a while, but sooner or later she wants to be caught.

Forced Matings

But when the buck is willing (99 percent of the time) and the doe is not (occasionally), I restrain the doe for a forced mating. I place her in the buck's hutch, rear end first, and hold her by the loose flesh over the shoulders (the scruff of the neck) with one hand and slide my other hand under her belly. I place thumb and forefinger on either side of the vulva, pushing gently toward the rear and lifting slightly. The buck will mount the doe, which is in a receptive position, and the service will take place. Sometimes, in the case of a reluctant pair, I have restrained the doe, picked up the buck, put him over her, and held her for him. If the buck acts reluctant and shy, keep trying. If you handle and pet your bucks regularly, they are less apt to be shy.

Technique for restraining the doe for forced mating.

Let me offer this final warning: Never leave the pair together unattended! They might fight and injure each other, and you'll never know if and when she is bred.

Is She Pregnant?

Just because a mating has taken place and just because they are rabbits, don't start counting babies until they are born. Conception may not have taken place. But how do you know for sure? There are several ways to tell, and one is more certain than the others.

First, however, weigh the doe at mating time and record the poundage. If later on she has gained a significant amount on the same feed that previously maintained her at a given figure, she probably is pregnant. If, for example, she weighs an additional pound in 2 weeks (for a medium-weight breed), she probably is pregnant. But you really can't be sure.

A week to 10 days after you have mated the pair, place the doe back in the buck's hutch to test her reaction. If she resists his advances and also growls and whines and generally complains during this test mating, it's a good bet she is pregnant. But it's only a good bet — it's not a certainty. If she weighs another pound, it's a better bet, but you still can't be sure.

On the other hand, let's suppose she does not resist the buck, or even if she does, you restrain her for a forced mating, and the buck completes a service. Record the date, because she may now be bred if she isn't already pregnant, and you will have saved yourself half the gestation period. We'll discuss records later.

Test-mating by putting the doe back in with the buck isn't conclusive because some does will complain and growl even if they aren't pregnant, and some won't even if they are. After a while, however, you will figure out (if you are observant) which does will complain when they are pregnant. Test-mating works for sure with some does, but you have to know which ones. It takes time and watchfulness.

How Do You Know for Sure?

The best and most certain way to tell if your doe is pregnant is palpation. Ten days to two weeks after mating, try feeling for the young. Pick the doe up and place her on a flat surface, perhaps on a feed sack or a piece of carpet on a table. Hold her by the scruff of the neck with one hand and slide the other under her belly. Feel around the sides of the belly, gently, for signs of life.

Within 10 to 14 days the youngsters on the way feel like large marbles on both sides of the center of the belly, just forward of the groin area. For comparison, feel another doe you know is not pregnant or even a buck. Palpation does take practice, but it's the only sure way to tell if a doe is pregnant.

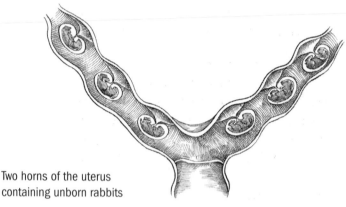

Two horns of the uterus
containing unborn rabbits

You're Still Not Certain?

Now you think you feel those large marbles. She has gained a pound. She growls like mad when you put her back in with the buck. Three-and-a-half weeks of the 28- to 34-day gestation period have passed. Throw a handful of straw on the hutch floor. If all the other tests have proved positive and she starts picking up the straw in her mouth and carrying it around, I'll bet you anything she is pregnant.

If the doe starts to carry straw around,
she probably will kindle in a day or so.

Time for the Nest Box

The gestation period is 28 to 34 days, but most litters are born on the 31st day. On the 27th day it's time to put in the nest box. You want the doe to have the box in time to get used to it and in time to build her nest in it but not too early. If it's there too soon, before she's overcome by the maternal nest-building impulse, she may use it for a bathroom. So wait until the 27th day after mating.

Earlier, in chapter 4, I pointed out the necessity for providing a nest box with an open top; refer to that chapter for suggestions on making or buying nest boxes. My favorite continues to be the wire-mesh box lined with cardboard in the winter and left unlined, except for the floor, when warm.

Warm-Weather Boxes

During the warm months, with overnight temperatures above 50°F (10°C), place about an inch (2.5 cm) of wood shavings in the bottom of the nest box. Shavings are good, but sawdust is not, as it can suffocate the young. On top of the shavings, place handfuls of straw to fill up the box. It should be nice, soft straw, not stemmy hay. Coarse hay often will be eaten and will not mix as well with the fur the doe pulls from her body to soften and insulate the nest. In extremely hot weather add only a little straw. The all-wire nest box is especially useful in torrid temperatures because of the ventilation it affords.

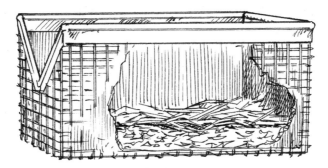

In warm weather, keep nesting material to a minimum. Too much may cause overheating of the litter. I use an inch (2.5 cm) of wood shavings, covered with a few handfuls of straw.

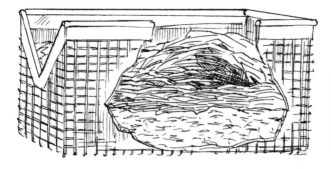

In cold weather, use several inches of shavings and fill the box completely with straw.

Cold-Weather Boxes

During the colder months, when temperatures fall below 50°F (10°C), line a wire nest box with corrugated cardboard. That is unnecessary with a wooden box, but a couple of layers in the bottom of a wooden box won't hurt, either. The cardboard provides insulation, should the young burrow to the bottom. Foam-plastic meat trays from the supermarket are also useful. The sheets of foam-plastic packing material that come with fragile merchandise also can be saved for this purpose.

I use 3 to 4 inches (7.6 to 10.2 cm) of shavings for additional absorbency and warmth. Then I pack in all the straw I can, cramming my fist way down deep inside to make a cavern I hope the doe will burrow into to kindle her young. She sometimes will, but even if she rearranges it all, she needs plenty of straw in the nest in the winter.

By the 31st day the doe should kindle the litter in the box. If it's an extra-large litter, she may kindle on the 30th day. The vast majority of my does, of all the breeds I have raised, have kindled on the 31st. Rarely will the litter arrive after the 31st day, but don't give up hope until after the 34th. After that, mate her again.

How the Doe Makes a Nest

On the day, or more often the night, the litter is born, the doe will burrow into the straw, pull fur from her underside, and build a nesting place for the young. After they are born, she will pull more fur and cover them up. If she does a good job, her litter will survive even in below-zero (–18°C) weather. If she doesn't, the litter may perish even in above-freezing temperatures. You can help her out, if she doesn't cover her babies well with fur, by plucking some from her yourself and covering the young. Or you can save fur from the summer months when there is more than enough and add it to the nest box in the winter. I have one doe who nearly denudes herself in July or August but is reluctant to part with her fur in the winter. So I save it for use when needed.

The New Litter

The doe needs peace and quiet a few days before the litter is born and a few days after. Dogs and children can be particularly disturbing at this time. Upsetting the rabbitry routine can cause the doe to kill her young or abandon them, so it is vital that you keep things calm and quiet.

You will, of course, be very curious to see the litter. If you have placed the nest box in the back of the hutch but in full view from the front, you will be able to see into it, and by the 31st day you should see a pile of fluffy fur toward the rear of it, moving slightly up and down.

Ten-day old babies in the nest. Newborns should be handled rarely if at all, so the doe will not be upset by the intrusion in her nest and so you avoid passing your scent to the young.

Time for the Tidbit

Now's the time to bring out that lettuce leaf, apple slice, half a carrot, or other such green tidbit. The doe will really appreciate it at this time. If you put it in the hutch, she will be so intent upon it that you will be able to remove the nest box with little or no opposition. Take it out of her sight and carefully push the fur aside. I use a wooden dowel to avoid touching the fur with my fingers and leaving a scent. If you can count the babies without picking them up, so much the better, but it's okay to move them aside to add up the tally.

Remove any dead ones, and if there are more than eight, you may want to foster the extras off to another doe who has fewer than eight, which is how many nipples a doe has. If there are any runts, those much smaller than the rest, you may want to dispose of them, although I always let them live. Sometimes they die in a few days, but sometimes they live and turn out just fine.

Foster Moms

If you have bred two or more does at the same time, which is a good practice, you can foster young from a large litter to a doe with a smaller litter. If you have different-colored rabbits, it's easy to record the color of the young that gain a foster mom. If all are the same color, mark the ear of the transferred youngster(s) with some ink from your tattoo kit (see Ear Identification later in this chapter). Just rub ink on with your finger. It will remain for several weeks. If it starts to fade, you can rub on more until it's time to tattoo a permanent number. It's important to keep track of any fostered babies.

Most does won't mind accepting young from other mothers, and if the young in both litters are within a few days of the same age, they will do nicely. But don't put a brand-new baby in with older and larger bunnies because it will never survive in the fight for the nipple. If a doe is nervous, or you simply aren't sure whether she will accept another youngster, rub a little vanilla extract on her nose, which will make it difficult for her to smell anything else. This smell will linger long enough for the newcomer to pick up the scent of the rest of the litter, and that's all it takes to gain social acceptance in the nest box. But it's really not necessary. Does adopt foster children quite readily. Attempts to hand-nurse orphans are usually futile because rabbit milk is much richer than any veterinary formulas you can obtain, such as those for kittens and puppies. Fostering to another doe is, on the other hand, quite successful.

Resist attempts to sex newborn and very young rabbits. It is difficult to tell the difference at this age, and it doesn't make any difference to me anyway. It is easy to sex them a few weeks later. The differences when rabbits are a few weeks old are apparent to anyone able to read this book. Besides, you won't be separating them until at least weaning time. Overall, the less handling of newborns, the better.

Extra Feed for the Doe

With the new litter looking for milk from the doe, it's a good idea to give her some extra feed. Of course, you will want her to have all the pellets she can eat; but she also will appreciate a lettuce leaf or some other green tidbit each day. If you have extra cow or goat milk, the doe and her growing litter will thrive if you can spare them some as a supplement. Dried milk is also good for the doe.

Out Comes the Litter

In about 10 days the litter will open their eyes, and after about 16 days they will come springing out of the box. In the meantime keep a daily watch on the litter in the box, make sure that all the babies are together in one place in the nest, and remove any that die. The doe has been getting a green tidbit now and then, but when the litter pops out of the box, the greens cease. You can remove the nest box when the litter is 3 weeks old in warm weather and 4 weeks old in winter.

Breeding the Doe Again

For years, normal practice was to wean the litter at 8 weeks and breed the doe again at that time, thereby gaining a maximum of four litters per year per doe. But since the nutritional advances that came with the manufacture of rabbit pellets, it is possible to maintain the doe in good condition even if you breed her again sooner than 8 weeks after kindling.

Ordinarily, my own practice is to rebreed the doe when the litter is 4 or 5 weeks old and wean the young at 6 or 7 weeks. That gives her 1 to 2 weeks without the litter to regain her strength for the next litter. And that gives me the potential for about five or six litters per year per doe. Some commercial breeders accelerate the breeding program even more, to seven or eight litters, by rebreeding the doe sooner and weaning the litter sooner.

Ordinarily, you will find it easier to get the doe to conceive if you breed her again while the litter is still running around the hutch with her. If you give her a rest between weaning one litter and breeding again, she is apt to gain internal fat that will stop her from conceiving. It isn't necessary to wean the litter at 8 weeks if you breed the doe again at that time; the litter may stay with her another couple of weeks. If she isn't bred again, the young may stay until they don't seem to be getting along — usually at about 3 months. You may leave young does with their mother almost indefinitely. Mothers and daughters get along just fine for months.

Weaning and Housing the Litter

Don't remove all the youngsters at the same time, because you want the doe to dry up gradually. Take out the biggest and huskiest of the youngsters; let the little ones stay in for some more milk. If you wean the litter over a period of about a week, the gradual process will be better for both mother and young. You may keep the litter in one hutch for another month (up to age 3 months), but after that time you will have to give each buck his own hutch, and it's best to have a hutch for each young doe, although two to a hutch is a satisfactory arrangement. If necessary, you can keep several does in one hutch, because they will continue to get along, but each buck needs his own or battles will ensue. All does to be mated need their own hutches.

If you are raising rabbits for breeding or show stock, it's a good idea to have several growing pens. These are smaller hutches, about 2 feet (0.2 m) square for the medium and smaller breeds. Both young bucks and does will do nicely in these, as you develop them for future breeding careers, for sale as breeders, or for show stock.

If you are raising rabbits for meat, however, you don't need growing pens (unless you like roaster-size eating rabbits; in that case, you will need some growing pens for them). You will find it most economical to slaughter or market medium-size meat rabbits at 8 weeks or so, because the weight they gain after that age is obtained on more pellets than the earlier poundage. Feed efficiency suffers. Most processors want to buy rabbits at 4 to 5 pounds (1.8 to 2.3 kg).

It is difficult to measure a rabbit's worth at 8 weeks, especially when it comes to body type. Former ARBA president Oren Reynolds always said that a rabbit has "all the rear end it will ever have the day it is born," but it takes a while to see how the shoulders will develop. If you are raising replacement or additional breeders, you will simply have to develop your rabbits for at least another month or so, depending upon the breed, before you can decide whether they should be kept or put into the pot. One reason that I like the smaller breeds is that they are still fryer-broiler size at 3 to 4 months of age. If I decide they are not worth keeping or selling as breeders, they still cook up very nicely, just like an eight-week-old medium-breed fryer.

The Problems of Weather

What about breeding rabbits in the very hot or very cold months? Years ago many rabbit breeders feared they would lose baby rabbits to hot weather, and some even constructed "cooling baskets" of wire screening in which they would place the babies during the warm days so they would have plenty of ventilation. That was before the era of the wire hutch and the wire nest box. Use of these two items makes the old cooling basket unnecessary and allows breeding right through the hottest part of summer.

The same equipment helps keep the litter warm in winter. Electric heating pads, encased in sheet metal, are made to fit the bottoms of the all-wire nest boxes and may be purchased very economically. If they save just one litter, they pay for themselves.

Many breeders take advantage of the wire top of the hutch, provided the hutch is under a roof, and place an aluminum photo-reflector light

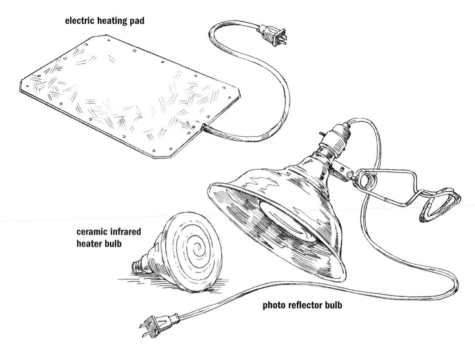

electric heating pad

ceramic infrared
heater bulb

photo reflector bulb

These devices help warm a litter in cold winter months.

over the nest box, so that heat from a bulb radiates down into the hutch. Others place such a light under the wire floor of the hutch below the nest box, and the litter burrows down into warmth. Ceramic infrared heater bulbs, sold in pet stores for reptile enclosures, also work well. Such practices really are necessary only on the day or night of kindling, or perhaps for a day or two after in really cold weather. By that time the young begin to gain some fur and can stand the cold.

Winter or summer, ventilation is always necessary, and the wire hutch and the open nest box provide it. A wooden hutch is often too hot for summer, and a closed nest box can become damp. Good ventilation is vital summer or winter. With today's equipment you can breed all year long.

Keeping Breeding Records

You simply must know when your does are bred. Otherwise, you'll never know when to install the nest box or expect the litter. That could lead to disaster. So you must record the dates and other information.

Doe Hutch Card

The illustration below shows what is required in the doe hutch card. It includes her birth date, name, and ear-identification number, as well as the names of her sire and dam. With this information you know when she's old enough to breed and which buck to select for mating if you have more than one available.

When you mate the doe, you write in the name of the buck under SERVED BY and mark the date of service. If you test mate her in a week or two and she mates, you must record the additional date, but respect the first one, and give her the nest box the 27th day thereafter. If she does not kindle on time, you can remove the nest box, but replace it 27 days after the second mating date. The rest of the items on the doe hutch card are self-explanatory.

The doe's hutch card is the single most important record. Faithful recording will reward you with valuable information that will help you decide how you will make further matings. If the record shows a nice healthy large litter resulting from the mating of that doe to the indicated

DOE HUTCH RECORD CARD

NORMAL WEIGHT ___

DOE ___ EAR TATTOO NO. ___ BREED ___ DAM ___

NO. NIPPLES ___ BORN ___ SIRE ___

BUCK	Date of Service	Pregnancy Check	Date Kindled	Kindled	Survived	No. Left	Weight	Remarks

(NUMBER OF YOUNG; AT 8 WEEKS)

STUD RECORD CARD

BREED ___ EAR NO. ___ REG. NO. ___ BORN ___
SIRE ___ REG. NO. ___ DAM ___ REG. NO. ___

DOE SERVED	Date of Service	Date Palpated	Date Kindled	Kindled	Survived	21 days	56 days	21 days	56 days	Exc.	Avg.	Poor

(NUMBER OF YOUNG; WT. OF LITTER; QUALITY OF LITTER)

buck, then you will probably want to mate the pair again. If not, try another buck or get rid of the doe.

Buck Hutch Card

Also significant, however, is the buck hutch card, especially if you have a number of bucks in your rabbitry. The information on the card will help you evaluate the buck in the future. Both doe and buck cards may be obtained from equipment dealers, or you may prefer to draw up your own, perhaps including space for additional information that you find useful.

I keep all the hutch cards required over the course of a breeding career right on the hutch so I don't need to transfer this information to other record sheets. I keep doe cards on the feeder for the life of the doe, stacking them on top of each other under the plastic holder. I can check production right on the hutch. The buck card has pedigree information on the back. I check the doe cards to see how the bucks are doing; that is, how many offspring they are siring.

Litter Production Records

Some raisers, particularly large commercial operators, maintain litter production records for all their bucks and does, and many computerize them. I find them unnecessary for the backyard raiser who keeps hutch cards and refers to them regularly.

Pedigree and Registration Certificates

The pedigree and registration certificates provide important information about ancestry. You should maintain the pedigree records of all your rabbits. If you keep or sell a young rabbit as a breeder, it must have a pedigree certificate. You can easily write one out from the information contained on the pedigree or registration certificate of its sire and dam.

I use the ARBA pedigree form, and I make and keep a copy for every rabbit I sell. I date the pedigree and file by date, the most recent first. If a customer loses a pedigree, I can replace it. I can always tell what additional rabbits would mate best with his. And if the customer wins best in show, I know which pair to mate again. The pedigree copies also help me supply a suitable rabbit when a satisfied past customer comes back for more.

Stock Record Book

Still another handy record item, of particular use to the raiser who sells breeding stock, is the stock record book. It can be simply a composition book or loose-leaf notebook. When rabbits in the herd are weaned, record breed, variety, color, name, ear-identification number, birth date, names and identification numbers of sire and dam, and any other facts you might want to refer to, such as weight at certain ages or quality of fur and markings. In my stock record book, I enter remarks such as "butcher" or "keep" or "good belly color" or "watch shoulders." I also rank each litter in the stock record book, with "1" being the best at weaning time. My ear-identification number system helps with visual record keeping too. You'll find details on this system in the next section of this chapter.

You may also want to enter the dates your rabbits were sold in your stock record book, as well as the names and addresses of the buyers. Keeping a list of customers helps in future selling. Or you may want to keep a separate sales record book. Increasingly, computers play a role.

Sales Record Book

In my sales record book, I record the dates of shipping or sale; names, addresses, and phone numbers of customers; and ear-identification number, date of birth, sire, dam, breed, variety, sex, and price of each rabbit sold. With this information at hand, I can also make out the pedigree.

IN THE FILES

I file all show reports, cards, leg certificates, registration, and grand championship certificates to assist when making out pedigrees. This way, I know what I have, what I have sold, what I should keep, and what to sell next. One improvement would be profiles of my customers: sizes of their rabbitries, their breeds, goals, and so on. These days, of course, you can keep all this information on a personal computer.

Ear Identification

Every rabbit above weaning age in your herd should have a permanent ear number tattooed in its left ear. Not only do you need to number the rabbits for your own identification purposes, but the permanent number helps to protect you from theft, marks the rabbit as one of yours when you sell it, and actually is required should you enter your rabbit in a show or register it with the ARBA.

With the plier-type tattoo set, you merely insert the numbers (or letters) into the pliers, squeeze firmly and quickly on the rabbit's left ear where the needles puncture the skin, release the pliers, and then rub the tattoo ink into the ear, working it in well with a brush or your finger. You can wipe away the excess, or just let it dry and flake off. I always wipe the ear first with a cotton ball dipped in rubbing alcohol.

As for identification numbers, you may use any letters and numbers you like. The tattoo pliers hold five letters or numbers. Some breeders put the rabbit's name in the ear, keeping it short, such as Kate, Jill, or Jack. Others number their rabbits in sequence from number one. I put my initials, BB, in the ear, followed by three numbers that tell me which month the rabbit was born, how many were born so far that month, and the year. If I number a rabbit BB139, it stands for Bob Bennett, born January, third rabbit marked that month, year 2009. (I use a zero for a rabbit born in October, an "A" for the month of November, and "B" for December.) After 10 rabbits

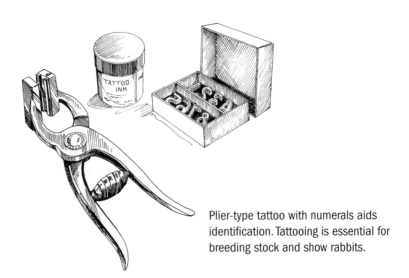

Plier-type tattoo with numerals aids identification. Tattooing is essential for breeding stock and show rabbits.

are born in a month, I use the numerals plus the alphabet to number the weaners. For example, BBAE9 is the 15th rabbit marked born in November 2009. The "E" is representative of 10 rabbits plus 5 (E being the fifth letter of the alphabet). Some people can tell a horse's age by looking at its teeth. I can tell my rabbits' ages by looking at their ears.

Use any system of coding that you find useful, but mark each rabbit, although you don't have to mark meat rabbits if you don't want to. Some breeders put letters MEAT or CULL in meat rabbits' ears, which would discourage another breeder from using these rabbits in a breeding program should they for some reason not find their way to the butcher.

HOW TO TATTOO AN EAR NUMBER

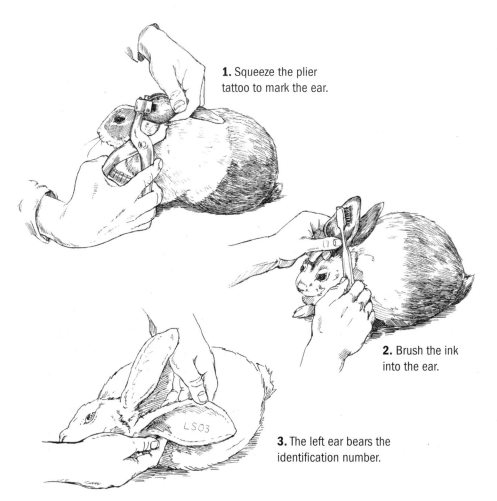

1. Squeeze the plier tattoo to mark the ear.

2. Brush the ink into the ear.

3. The left ear bears the identification number.

WHY THE LEFT EAR?
The tattoo number goes into the left ear because the right is reserved for the registration number, which the rabbit may earn at maturity if it passes the registrar's examination.

Choosing Future Breeders

How do you decide which rabbits to keep for your breeding program, or for the shows, or to sell as breeders? Many books have been written on the subject of animal breeding and genetics, and the principles apply to rabbits as well as other animals.

Selective Breeding

There are several rabbit-breeding systems, but selection lies at the heart of all of them. *Selection* is the choosing, the selecting, of those youngsters that you think will make the best future breeders. How well you choose will determine whether you are a rabbit breeder or just a rabbit raiser. Breeders do more than merely propagate rabbit species. They do that, to be sure, but in such a way that they maintain and even improve the high quality of their rabbits, which consequently get better and better generation by generation.

What Is Better?

Better depends upon the breeder's goals. For the meat producer better means greater meat qualities. For the Angora raiser it means more and longer wool. For the fancier it means improved body type, color, markings, and so on. But for all breeders it means better health, greater feed efficiency, and larger litters, among other things.

Here are some guidelines: First, make sure you like the parents. The youngsters you are considering should look as good as, and preferably better than, their parents. The dam must be easy to breed. She must conceive consistently. She must have large litters. She must maintain perfect health. The sire must be a good breeding buck of good body type, have good fur, be healthy, and sire large litters. Both must be of good temperament (although a rabbit with a bad temperament is a rarity). There is a saying, "The young prove the parents," meaning that a pair that produce outstanding offspring certainly are worth mating repeatedly. You don't know, of course, if the offspring will be truly outstanding until you see how they perform as breeders themselves. The great thing with rabbits is that it doesn't take very long to find out. Rabbits reach breeding age in a matter of months, but their breeding careers last for several years.

The Meat Producer's Perspective

If you are raising meat rabbits, save youngsters that have good shoulders, loins, and rumps; gain weight rapidly; and grow faster than their littermates. You must concern yourself with production goals and body type. In the case of New Zealands, for example, blocky bucks and longer-bodied does will give you the meaty fryer-broilers you want to produce.

Good body type is important to the rabbit meat producer.

The Fancier's Perspective

The fancier, in addition to selecting for all the same factors as the meat producer, must also choose future breeders expected to score well against the point system carried in the standard for that breed. In the case of Tans, for example, color and markings make up much of the point scoring in that breed standard. So in addition to being strong, healthy producers, Tans must have certain color factors and markings. The goal of the fancier is to produce a superior rabbit, one that has great body condition, produces excellent fur or wool, and breeds superbly. That's why most rabbit raisers, no matter why they raise them, look to the fanciers for foundation stock.

Keep Your Goals in Mind

When making decisions about selection, you have to keep your goals in mind and choose as future breeders those that come closest to meeting these goals. Don't expect to breed the perfect rabbit, but view each specimen as half of a breeding team. For example, a weak-shouldered doe may be okay if you have a burly-shouldered buck. But don't choose two rabbits with the same failing, or you may breed in the weakness.

Rating sheets take a little time to complete, but because Tans (see Raiser's Edge: A Model Rating Sheet) are at their best between 6 months and a year, you have to rate them only once.

The designer of the rating sheet shown in the Raiser's Edge box below encourages Tan breeders, in particular, to use rating sheets, because with dominant and recessive *alleles* (alternative forms of a gene) involved in creating the ideal Tan, the sheets make it easier to:

- Identify which animals should be mated with which
- Mate animals that do not have the same weaknesses
- Mate animals strong on points you need or want to intensify in your herd
- Improve overall quality of your breeders by eliminating animals with low-average or poor ratings
- Create a progeny-test record: by sorting ratings by dam or sire, you can see how well each is transmitting desirable characteristics

---

R A I S E R ' S   E D G E

A Model Rating Sheet

CLAY STEINBERGER of Connecticut raised superior Tan rabbits for more than 20 years, and he did it scientifically. He sent me a copy of a rating sheet he developed to help him breed better Tans. Using the same approach and making revisions to suit your purposes, it could help you improve any breed of rabbit.

"Enclosed is a rating sheet," Clay told me. "I made it up as a tool to improve my chances by giving me a paper basis for determining matings. I don't have to tell you that pedigrees don't give you any help in breeding, but knowing what qualities each animal has — how persistent these are in a line, what qualities are being transmitted in offspring — helps determine which animals should be your breeders."

With Clay's permission, I share his rating sheet with you (at right). *Revise it to suit your breed* (with spots, blazes, whatever).

TAN RATING SHEET

Date _____

Name _____   Sire _____

Ear # _____ Born _____   Ear # _____

Sex _____ Weight _____   Dam _____

Ear # _____

Ratings: 1 = excellent, 2 = good, 3 = average, 4 = low-average, 5 = poor

Body Type	Rating	Markings (cont'd)	Rating
Compactness	1 2 3 4 5	Triangle shape	1 2 3 4 5
Shoulders	1 2 3 4 5	Color intensity	1 2 3 4 5
Rump	1 2 3 4 5	Ears	
Tightness	1 2 3 4 5	Substance, size	1 2 3 4 5
		Color intensity	1 2 3 4 5
Fur		Crown spots	1 2 3 4 5
Fly back	1 2 3 4 5	Chest	
Length	1 2 3 4 5	Width	1 2 3 4 5
Luster	1 2 3 4 5	Color intensity	1 2 3 4 5
		Cleanness	1 2 3 4 5
Color		Flanks	
Body color	1 2 3 4 5	Straightness	1 2 3 4 5
Depth (body color)	1 2 3 4 5	Line placement	1 2 3 4 5
Tan (richness/redness)	1 2 3 4 5	Cleanness	1 2 3 4 5
Depth (tan color)	1 2 3 4 5	Belly	
Evenness (tan)	1 2 3 4 5	Color evenness	1 2 3 4 5
		Color intensity	1 2 3 4 5
Markings		Tail (color intensity)	1 2 3 4 5
Cleanness (overall)	1 2 3 4 5	Brindling	
Nose	1 2 3 4 5	Rump	1 2 3 4 5
Nostril	1 2 3 4 5	Sides	1 2 3 4 5
Upper lip	1 2 3 4 5	Feet and legs	
Jowl (color intensity)	1 2 3 4 5	Front (cleanness)	1 2 3 4 5
Eyes		Rear (placement)	1 2 3 4 5
Eye-circle size	1 2 3 4 5	Line straightness	1 2 3 4 5
Eye color intensity	1 2 3 4 5	Color intensity	1 2 3 4 5
Collar	1 2 3 4 5	Toenails	1 2 3 4 5

Faults: _____

Comments: _____

_____

_____

- Sort by breeding line or "families"
- Give you a record of which mated pairs *nick* — that is, produce a high percentage of good youngsters
- Train you to observe *all* qualities to be considered in breeding and improve your ability to judge the better animals
- Computerize the whole operation

Methods of Breeding

That brings us to the subject of breeding systems. Should you inbreed, line-breed, outcross, or crossbreed? First of all, we had better define our terms.

Inbreeding is the breeding of close relatives, such as brother and sister, mother and son, cousins, and so on. *Linebreeding* is a form of inbreeding that follows a line of descent, usually from an outstanding ancestor, and involves the use of relatives such as grandmother and grandson, aunt and nephew, grandsire and granddaughter. Most successful rabbit breeders use a form of linebreeding. In other words, they inbreed on a line of descent from an outstanding rabbit or rabbits they feel will maintain good quality and improve their stock.

Outcrossing is the use of an unrelated animal of the same breed in the breeding program. Most breeders resort to an outcross at one time or another, no matter how much they would like to keep it in the family. *Crossbreeding* is, of course, the mating of two different breeds. Some successful breeders of meat rabbits have raised pure strains of New Zealand Whites and Californians and then made judicious crosses to produce a hybrid fryer. Some Angora wool producers have crossed different Angora breeds with good results. Generally speaking, however, crossbreeding is better left to the experts or those of an experimental turn of mind. The following experience shows why.

For four years I experimented with a cross between the Tan and the Netherland Dwarf, in an attempt to produce a Tan Dwarf, of which there were none to be found (although there was a standard for them). This effort took me through many generations, and I only got close; I finally gave up. So I can tell you that it is a time- and hutch-consuming process with very little reward. And I was only trying to produce a variety already written into the standards, not a new breed or variety. Today there are Tan Dwarfs

WHEN CROSSBREEDING
IS A GOOD THING

Farmers who raise livestock such as cattle, hogs, sheep, and poultry for a living have long relied on crossbreeding to obtain the best possible stock. Farmers begin with a crossbred sire that is the product of two or more distinct purebred breeds and a crossbred dam that is the product of other distinct purebred breeds. The four (or more) breeds used to produce the sire and dam are selected for their various desirable attributes, such as size, hardiness, good feed conversion, and mothering ability. When the sires and dams these four breeds produce are mated, the resulting offspring exhibit the best qualities that the several breeds have to offer. We call such offspring *hybrids*, the result of a four-way (or more) cross.

Because the principles of genetics apply to plants as well as to animals, many of us already have a rudimentary understanding of hybrid crosses. Consider hybrid corn, tomatoes, petunias, and impatiens, for example. If we save seeds from these hybrid plants, often they will not "breed true" (that is, come out looking like the plants from which they came) when planted.

Likewise, in the development of hybrid animals, it makes little sense to save offspring for continued breeding. Inevitably, we would lose the positive traits that we sought in the first place. To gain replacement sires and dams, we must return to the source.

As you might suspect, hybrid rabbits have been developed over the years as well. The English were the first to popularize a hybrid breed, but the Altex is a commercial-sire breed whose name pays tribute to the states where it was developed — Alabama and Texas. The breeding project began in the 1980s with purebred Californians, Champagne D'Argents, and Flemish Giants. The result was a 13-pound buck typically mated to a New Zealand doe or to a doe that is a cross between a Californian and a New Zealand.

Many producers report that their crossbred fryers reach market size about a week earlier than purebred New Zealand fryers. The dressout percentage and the meat-to-bone ratios of the crossbred fryers are also better. We can only imagine what the future will hold for animal genetics if cloning becomes a practical reality for livestock producers.

in existence, so somebody was successful in a task that really was beyond my will to pursue. New breeds and varieties continue to come along, but improvement of existing breeds is a goal worthy enough for me.

Why Inbreeding?

Most of us recoil from the thought of inbreeding, the mating of close relatives. For humans incest is taboo, and law forbids marriage of close relatives. When it happens anyway, the offspring are riddled with problems. So how can inbreeding be any good for rabbits?

Inbreeding accentuates existing characteristics, both good and bad. If you mate two good animals, you can produce something better. If you mate two bad ones, the results can be worse. Humans themselves proved it a long time ago.

Choose Excellent Foundation Stock

When mating close rabbit relatives, you have to select for breeding only those that have the characteristics you want in the offspring.

You must begin with excellent foundation stock and know your breeding goals. You can then inbreed on a line of descent that will guarantee a continuing parade of fine rabbits.

Inbreeding brings no new blood to a rabbitry. If you don't have good foundation stock, you won't get very far by inbreeding; you'd be better off introducing new blood. In fact, you'd be better off introducing 100 percent new blood by getting other rabbits. When potential rabbit breeders start out with inferior rabbits, they learn what's lacking, such as size, body type, or color, and then add new blood in the hope that it will make up for the deficiencies. The theory is that at some point they will have the rabbits they want and then will begin inbreeding. In practice that's what happens for most people. But it is far better to start with superior stock and then try to limit the gene pool or maintain that superiority on a line to future generations.

Where Two Pairs Can Take You

Let's assume you have begun with two pairs of at least distantly related rabbits from the same breeder. You breed the two pairs, and two litters result. If you want to increase the size of the herd, you will want to breed the two pairs again, meanwhile saving offspring from the first two litters. If you save three does and two bucks from each litter, you might:

- Mate a buck from one litter with a doe from the other
- Mate a doe from each of the two litters back to her sire
- Mate a doe from each of the two litters to the other doe's sire
- Mate a buck back to its dam
- Mate the same buck to the other litter's dam

The possible results of this activity, over a period of several months, would be nine litters.

Selecting Future Breeders

The nine litters produced from the above scenarios would provide a pool of talent from which to select future breeders. Here's how to select specific rabbits for future breeding: Each prospective new breeder should at least equal, and preferably exceed, its sire and dam in selection qualities. Of course, that sire and dam should possess the desired characteristics, as determined by yourself or their former owner. And so it can be seen again how important it is to begin with outstanding foundation stock, and why you should rely on the judgment of a respected breeder to start you off right.

If in your breeding program you continue to select outstanding offspring from outstanding parents and consign those that do not measure up directly to your pot, freezer, or a meat processor, you will be on your way to building a successful herd. At the same time you must choose, on a line of succession, those offspring and forebears to mate together. No two mates should possess the same weakness, or the resultant litter will possess a double dose of that weakness and will be a disappointment. But if you select wisely, the litter will receive a double dose of good qualities and only a single dose of a weakness, and you will be delighted with the litter.

Of course, some individual rabbits will always be better than others. It is up to you to recognize what better is in your own terms and ingrain it into your herd by mating those better animals. You will be able to do that by careful study of your animals and the breed standard and by keeping your breeding goals in mind. Examination by a registrar and comparison of your animals by judges at shows certainly will help. But the biggest responsibility lies with you, because most of the really important qualities of your stock will be found only in the breeding hutches and the nest boxes. Others' opinions are welcome, but you should always be your own best judge. That requires experience and study, and it also means you want your rabbits to please yourself, above all.

Outcrossing

Perhaps, however, your foundation stock was not all it should have been, or your first attempts at selection took your stock downhill instead of up. All your best efforts at line breeding failed to produce the desired results. You may want to resort to an outcross—within your breed and variety, of course. Almost every breeder does so at one time or another. I have used imported animals in hopes of improving my own when a characteristic mine lacked was found abroad. As a rule, however, it is wise to avoid outcrossing if possible, because the introduction of new genes is more likely to confuse the situation than help it. You can't be as sure of what you will get (unless you are familiar with the forebears of the foreign or unrelated rabbit).

Reminders That Bear Restating

Start with the best possible foundation stock. Mate the best offspring back to the best parents, and inbreed this way on a line of descent from the best. Resort to an outcross sparingly. Remember that the offspring are proof of the parents' ability to produce. Keep breeding only those parents that produce the kind of young you want. That might take a few litters to determine. You don't necessarily want to keep breeding the first pair. You've got to find out if they are producing what you are looking for. Improvement is a slow process, but it is steady if you follow the rules of good sense and read up on genetics.

Once you think you have bred the perfect rabbit, you might as well find something else to do with your time. The perfect rabbit never appears, however, because whenever anyone gets close, the standard is rewritten to make perfection just a little harder to attain. And what a pity if we were able to achieve perfection, for all the fun of pursuit of it would suddenly disappear.

PART III

Marketing and Miscellany

7

To the Best Market: Selling Rabbits

IF YOU START RAISING RABBITS RIGHT, you doubtless will sell some whether you really want to or not. You'll have nice rabbits, and people will want to buy them. They will wave the money at you, and you will, if you are like most of us, take it.

They will want your rabbits for meat, laboratory use, breeding stock, or pets. And even if you have only a very few breeders and a voracious appetite for rabbit, you will have extra rabbits to sell.

Of course, success in selling rabbits depends upon, as pointed out earlier, choice of a breed for which there is some kind of accessible market already. Let's look a little closer at some of these markets.

The Meat Market

Because many Americans haven't yet discovered the nutritional benefits and the good eating rabbit meat offers, the meat market for rabbits in the United States has tremendous potential for growth.

Live Meat Rabbits

The easiest way to sell meat rabbits is to sell them live to a processor. Some processors pick them up at your rabbitry. Others want you to deliver. Either way, you have to find out how many and what size rabbits the processor wants and when. You can locate a processor by talking to other breeders in your area. Personnel at your feed store may be able to identify one, so be sure to ask them. Processors often attend local club meetings, which is just one of the benefits of club membership.

While selling live to a processor is the easiest way, it is not necessarily the most lucrative, particularly for the small raiser. The price you receive will be the lowest possible return you can get for your rabbits. Sell live meat rabbits to a processor only if you have neither the time nor the inclination to make a greater effort for a correspondingly greater return.

You may, for example, be able to gain a better price by selling live to a meat market that will also do the butchering and thus cut out the processor's profit. The butchers will be able to make a greater margin and pay you a little more. The same may be true of a restaurant or an institution, such as a hospital or a school.

You may also be able to develop a list of customers who will buy rabbits live for their own tables. Many rabbit raisers have acquired a waiting market just by starting with live sales to friends, relatives, neighbors, and coworkers. These people will pay you a little more than will the processor, but they'll still be getting meat for their tables at the lowest possible price short of raising rabbits themselves.

Dressed Meat

If you have only a few rabbits to sell for meat, or a lot of spare time, you will find more profit in selling dressed rather than live rabbits for meat. You may have to meet local requirements for slaughtering and butchering, which will require knowledge of health department regulations. Commercial butchering is not difficult but must be done under regulated sanitary conditions.

You may develop customers for dressed rabbit meat among relatives, friends, neighbors, and coworkers, as well as area restaurants, stores, and institutions. For this market, you may expect a return much greater than you can receive for live-meat sales, but of course you must have the proper facilities.

Slaughtering and Butchering

Home slaughtering and butchering for your own family's consumption are simple procedures and may be carried out in the garage or basement. Rabbit meat can be packaged in foam-plastic trays with plastic wrap. Or you may simply place it in freezer bags.

Slaughtering

Take the following steps to slaughter a rabbit:

1. To make the structure that will hold the carcass, drive some large nails into a thick board and hang it on a wall with a wire and another nail or whatever works. Hang the plastic garbage bag on the nails.

MATERIALS
- Knives
- Gloves
- Nails
- Plastic garbage bag

Step 2: Kill by snapping the head back, breaking the neck.

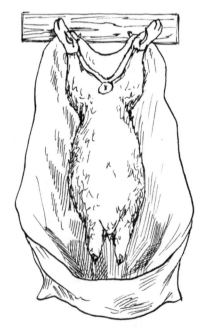

Step 3: Hang plastic bag, then carcass on nails or hooks for bleeding.

2. To kill the rabbit, hold it by the hind feet and quickly snap its head down and back, breaking the neck.

3. Cut off the head with a knife or hatchet and hang the carcass upside down by the hind feet on the nails over the bag, and let it bleed until it stops.

4. Cut the skin around each hock with a skinning or boning knife. Slit the skin from hock to anus inside each leg.

5. Pull the skin down over the carcass, as you'd peel off a pullover sweater. Put the pelt aside.

6. Slit the carcass from tail to abdomen, cutting around the anal opening between the hind legs. Save the heart, kidneys, and liver if you wish, but let the rest of the entrails fall into the plastic bag for disposal.

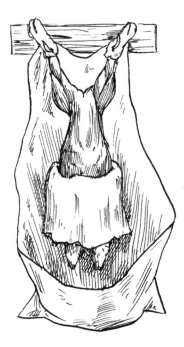

Step 5: Pull the skin down over the carcass.

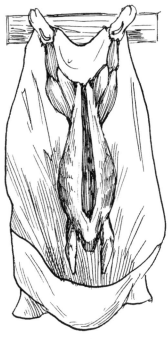

Step 6: Slit the carcass and let entrails fall into plastic bag.

Portable Abattoir

P HIL ANGELL of Dallas, Texas, devised a portable *abattoir*, or
small-animal dressing station. Phil simply took a plastic 55-gallon
drum, cut off the top, and sawed it in half lengthwise down to about
8 inches from the bottom. Then he bolted two pieces of 2 × 3 lumber
to the outsides to make vertical uprights and added a 2 × 3 crosspiece
about 6 inches above the top of the barrel. Skinning hooks, available
from feed stores, hang down from the crosspiece, and there you have it.
Phil uses his on the tailgate of a pickup or on a table because he likes to
stand when he dresses a fryer. But you could put it on the ground and sit
on a stool to use it.

Phil's abattoir is just right when you want to dress a lot of rabbits. If
you are dressing just the occasional animal or are trying it for the first
time, you might want to use my plastic-bag technique pictured on the
preceding pages.

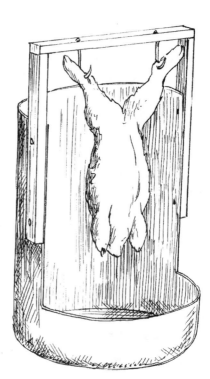

Use a 55-gallon (208 L)
drum to make a handy
rabbit-dressing station.

Butchering a Rabbit

After you've slaughtered the rabbit, run the carcass under cold running water like you would any cut of meat you might get from the supermarket. Then refrigerate the carcass (in plastic wrap or on a dish or tray) overnight or for at least a few hours before butchering. It is easier to make clean cuts in chilled meat.

MATERIALS
- Knives
- Cutting board
- Plastic wrap, freezer bags, or plastic-foam trays

1. On a cutting board, split the hind legs from the back.

2. Cut the back into two or three pieces, depending upon the size of the rabbit.

3. Cut the forelegs apart.

4. Wash all the pieces under running water, removing any hair that might cling to them.

5. Package as desired. I don't sell any dressed rabbit meat, so I don't bother with fancy packaging, which you would want to consider if you were offering the meat for sale.

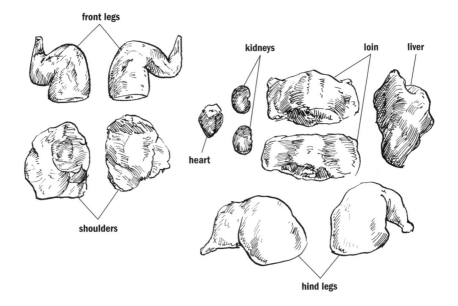

What about the Pelt?

A possible additional source of income from both dressed-meat sales and rabbits raised for home consumption is the pelt. It should be dried on a spring-wire shaper, with the fat first scraped from it. Several pelts, once dry, store nicely in a plastic bag, with mothballs. You may pay to have them tanned, tan them yourself, or simply sell them dried. Prices paid for dried pelts fluctuate and never seem to amount to much if the pelts are from young fryers. Perhaps you can locate a store that sells novelties and would like to carry some of the tanned pelts. Or you might know someone who makes stuffed animals and pillows. You or another member of your family can also try to make them into stuffed articles to have or sell.

Spring-wire
shaper for pelt

I have never tanned a rabbit skin because it looks like a time-consuming job, and I'm not going to recommend that you try it. I should add here that I get my pelts tanned for free. A friend of mine sends them away to be tanned and keeps half of them.

THEY SAY TIMING IS IMPORTANT

I've long been a proponent of rabbit shows on Easter Sunday because that's the time of year when many people have an interest in rabbits. As far as I know, however, nobody holds them then because all the rabbit raisers want to stay home and eat ham.

I learned a new way to take advantage of timing — specifically, during hunting season — when I saw the following classified ad run repeatedly in the local paper. It pays to market creatively.

> **Attention Rabbit Hunters:**
> Don't go home empty-handed. Try domestic rabbit meat. Delicious, nutritious, and versatile.

WOULD YOU BELIEVE, A RABBIT THAT LAYS EGGS?

I've always marveled at the versatility of the rabbit. Meat, wool, fur, research subject, fertilizer, fun at shows, pets, small business, big hobby, occupational therapy for young and old: the list goes on.

But what if a rabbit could lay eggs? Boy, that would really be something, wouldn't it?

I won't ask you to believe that my rabbits lay eggs. But I do have a friend who raises chickens and swaps me a dozen jumbo eggs a week for one dressed rabbit a month. I think it's a terrific deal, and so does she.

The Laboratory Market

Obtaining a contract for direct sales to a laboratory is not the easiest thing a new raiser can do. You must meet strict standards of management and sanitation, but that's not the hard part. What's difficult is supplying the large number of laboratory rabbits that may be needed. The way most breeders start out, therefore, is to sell to a broker, ordinarily a larger breeder who supplements his own laboratory stock production by buying from others.

You can start by contacting hospitals, universities, and pharmaceutical companies in your area to obtain information about their requirements for laboratory stock. They may tell you that they're already obtaining laboratory stock from someone else. Nevertheless, leave your business card. You may yet get a sale or a referral. Also, contact the supplier they've mentioned, who may need additional rabbits. Another way to find out who is buying laboratory rabbits (and all other kinds) is through your local rabbit breeders association or club. (See chapter 11 for more on how associations can help.)

Breeding Stock

Sales of breeding stock will bring the greatest return per rabbit to your rabbitry. Breeding stock will be the best of your meat or laboratory production that is surplus to your own breeding requirements. It is the most difficult but the most lucrative way to sell your rabbits. It's not just the rate of return that's important: The more people who raise their own rabbits, the more rabbit meat will be available, and the sooner it will become a dietary staple.

When selling breeding stock, try to have at least two litters of young available at all times, because few people want to start breeding from a brother and sister pair. The uninitiated do not understand inbreeding, and even if they did, such a pair does not necessarily provide a good

RABBIT ON EVERY DINNER TABLE?

Way back in 1988, *Advertising Age* magazine, read each week by advertising professionals, reported that Ore-Best Farms — a USDA-approved rabbit-processing plant in Oregon City, Oregon, that shipped vacuum-packed rabbit fryers to Oregon supermarkets — was banking its expansion plans on rabbit showing up on American dinner tables as often as fish or lamb by the end of the decade.

With a large ad campaign that quite noticeably skirted any suggestion that rabbits make cute pets, the company hoped to tempt the American palate and raise consumers' consciousness of rabbit as a high-protein, low-fat meat — higher in protein and lower in cholesterol, calories, and fat than other meats, including chicken, and relatively inexpensive too. Why wouldn't Americans, who were checking cholesterol, counting calories, seeking alternatives to red meat, and increasingly focusing on health issues, find "tender country-raised rabbit" an attractive, affordable delicacy-turned-staple?

Because, as it turns out, most Americans still do think of rabbits as cute, cuddly pets and cannot get their minds or their mouths around rabbit as a food, as Europeans do so readily. So Ore-Best Farms' profit prophecy failed to materialize. If reeducation of the public is what it will take to make rabbit meat a dietary staple in the United States, then it would seem that most Americans are still in kindergarten.

foundation for a breeding program. It's always a good idea to breed two does at the same time anyway, and if you do, you will always have at least two litters from which to sell breeding stock.

As you gain experience with your breed, you will be able to tell at fryer-marketing time (about 8 weeks for the medium breeds) which rabbits are better than others. Hold these back from the meat market for prospective breeders. By age 3 months, for most breeds, you will be able to tell if these rabbits are in fact good potential breeding stock. I wouldn't sell any rabbits for breeding under this age except in the case of some marked breeds, such as Dutch, whose value lies in color and markings to a greater extent than others. These characteristics are apparent at an early age.

For most breeds you really can't make a valid judgment until the rabbits are at least 3 months old, so you will want to retain the young at least until that age. You can keep does together almost until breeding age, but bucks must have separate cages at about 3 months. Actually, however, that will not entail a great many hutches, because you will not need to retain as many bucks as does. Not only is there a greater demand for does initially, but the ongoing need is greater because does have a shorter breeding career than the bucks. Does will bear litters for about 3 years; bucks will sire them two to three times longer.

Advertising Your Breeding Stock

While a ready market may await your meat and laboratory production, you must create a breeding-stock market. One of the best ways to do this is by advertising.

The classified pages of your local newspaper will work hard for you. A small-town daily or a weekly paper is usually better than a large city daily. Throughout the United States there are a number of classified-ad newsletters that are published locally or regionally and charge only a percentage of sales as a commission. One in my area charges 10 percent, which is typical, and only after the sale is made. I am able to advertise at no cost unless and until I sell. And I build the price of the ad, the commission, into my selling price.

The local feed store where you purchase your rabbit pellets usually has a bulletin board where you may advertise your rabbits. Just an index card or a business card tacked on the board will start to bring you customers.

Bob Bennett

My Famous Tans

133 Governors Lane
Shelburne, VT 05482 (by appointment)

802-985-8597
rabbitalogue@hotmail.com

National advertising in magazines will also bring customers. Resort to national advertising only when you have a supply large and steady enough to make it pay. Don't advertise nationally if you have only a dozen or so rabbits to sell. Not only will they hardly pay for the ad, but they may already be sold locally by the time the ad appears in the magazine. Many rabbit raisers have Web sites and correspond with customers by e-mail. You may want your own site. Rabbit clubs also have them, and they list their member raisers. Web classified sites such as www.craigslist.com list rabbits for sale, but the last time I looked eBay would not. We all know the Web is a great place to buy and sell, so do some online research.

Small and Steady

Regardless of where you advertise, you will get your best results from small but frequent advertisements. Be a steady advertiser. Be known as someone who always has rabbits for sale, all year long. And be the steady producer who can keep on advertising. If you have a steady supply and advertise steadily, you will find a continuing demand for your stock.

If you advertise locally, such as in the classified pages, don't give out much information over the phone. Don't mention any prices. If you are pushed, give only a range of prices. Try to get the caller to come and see

R A I S E R ' S   E D G E

Customer Information a Big Priority

P AT LAMAR of Washington State describes herself as "Head Fuzzy" of the Finley Fuzzy Farm Rabbit Ranch, where for years she managed a herd of four hundred rabbits with some help from her husband and daughter. A key feature of her business was supplying information to her customers, which she accomplished primarily with two well-written publications that helped sell her New Zealands, Rex, Flemish Giants, and Mini Lops.

Pat prepared the booklet *Your Rabbit and You* specifically for new owners of pet rabbits. It contains excellent basic feeding, housing, and management instructions. In addition to helpful information on rabbit care, the booklet invites customers to call her with questions after 7 p.m. any night of the week.

The "Fuzzy Fables Newsletter for Our Valued Customers" was printed quarterly and helped Pat keep her customers up to date on the goings-on at the ranch and share issues of interest to rabbit owners.

your rabbits. Nice rabbits will do the selling for you. If you give prices over the phone, you may regret it. People are funny. If you tell them that a rabbit costs $10, some will think it's too much and some will think it's too little. When Netherland Dwarfs were selling at about $50 each as they made their debut on these shores, a friend of mine tried to sell some for $25 but had little success until he raised his prices. Buyers wanted to pay more because they felt they were purchasing a high-class item.

Find the Local Customer: Plan a Workshop

If you want to sell something — anything — you need a bona fide prospect who is likely to buy what you are selling. Professionals call them "qualified." The most qualified prospects already have rabbits and have already demonstrated interest. What they need, of course, is more and better rabbits — yours.

The second-most-qualified prospects already have animals — perhaps chickens, sheep, goats, or gamebirds but not rabbits. These prospects have demonstrated an interest in animals already. What they need, of course, are rabbits. Right again: yours.

There is one place to find both kinds of prospects. You go there all the time. Sure, it's the feed store. Other rabbit raisers and other animal raisers frequent the feed store for the same reason you do: They simply must. Make the feed store your primary local marketplace.

Cooperation from the feed-store owner or manager is required and easy to get. If you sell more rabbits, the store sells more feed, cages, and other supplies it should stock. Make that clear. Okay, the store sells more feed. But customers are asking more and more questions about rabbits. Either the owner or manager doesn't know the answers or hasn't got the time to talk. Here's where you become the expert. Yes, you are the local expert on rabbits. Oh, you don't have to say so. All you have to do is offer to answer the questions. Then the store personnel will say you are the expert.

Contact your feed dealer at a good time to talk. That's not on a Saturday morning. But Saturday morning is an excellent time for you to be the expert. Offer to be available on a busy Saturday morning, say from eight o'clock until noon, to put rabbits on display and answer customers' questions about care and feeding.

Plan the day about a month in advance. Your feed dealer may even be willing to put an ad in the paper with your name and a picture of you with some of your rabbits. You won't know if this is possible until you ask. Put posters up in the store to announce the event. Ask the dealer to write to all rabbit-feed customers, inviting rabbit raisers to come and see your rabbits and ask you questions. You help with the letter, but it goes out with the dealer's signature on the store letterhead. Get the store to keep names of rabbit-feed buyers if it doesn't already.

In the meantime you can start your own list of raisers by announcing a drawing for a free pair of rabbits and a bag of feed. Install a slotted box topped with an informational poster and a supply of entry tickets for the drawing at the store. Be sure to include space on the tickets for both name and address. Have the drawing on the great Saturday you will be there. One small dealer I know has 75 regular rabbit-feed customers who buy at least 50 pounds (22.7 kg) a week, but only 8 of them belong to the local rabbit club. Keep in mind that lots of people like rabbits but don't like clubs.

R A I S E R ' S   E D G E

Feed-Store Workshop Testimonial

JUST TO PROVE TO YOU that holding a workshop in a local feed store will garner customers for you, I offer the following letter I received from Pat Barberi, a fellow Vermonter.

"After giving a day like this considerable thought, I decided to give it a try. The manager liked the idea immediately and said he would help advertise, and we picked the day.

"On that Saturday I went down and set up. I brought a card table and set out brochures from the feed store, the American Rabbit Breeders Association, and copies of one of my favorite books on rabbits. I had several representatives of my breeds: Holland Lops, Mini Lops, New Zealands, and Californians. I also displayed feeders and crocks that were carried by the store.

"I had only a few small juniors for sale, so the manager obtained some from another local breeder. During the morning I sold all the young rabbits, a total of eight, plus a pair of senior Mini Lops with papers. In addition, I invited several people to visit my rabbitry (which resulted in additional sales later).

"I wore my lab coat, adorned with club patches, which caught people's attention as they came in and kept store employees from being asked too many nonstore questions.

"Most people were curious about the different breeds and their uses. I had my various rabbit books handy to look up information I didn't have memorized. Many of the people buying a rabbit also wanted to build a cage. It was very handy having the wire there at the store available to them, as well as directions in the books for sale.

"It was a very pleasant place to spend a day and a very rewarding experience, besides the money, talking to so many people. Many customers did not realize the extent of the products for sale at the store, such as the various types of grain that can be used for supplementing a rabbit's diet (whole oats, Calf Manna, sunflower seeds, and so on). It was also helpful to have medications there when discussing the health problems rabbits can have.

"Aside from the obvious benefits I received in promoting my own stock, it was a good feeling to know that these rabbits were off to a good start with their new families. I am certainly planning to return, as the store management also considered the workshop worthwhile."

Pat reported later that the store bought a newspaper ad and was referring potential customers to her.

Make arrangements ahead of time for the placement of your hutches. Near the front door or the loading dock are good spots. Put your rabbits in clean wire hutches, preferably the type your feed dealer sells. Move some of the dealer's hutches and other equipment next to your rabbits. Books for sale and free literature would add to the display.

Put up a sign that describes your breed and lists your name, address, and phone number. Hand out a brochure, business card, or sales letter. If you don't have a letter, type or print one and photocopy it. Answer questions. Raisers who have only mongrels or some other breed may become interested in your breed. Invite them to your rabbitry. Write down everyone's name, address, and phone number. If they don't call you, then you call them.

After the big day, make sure your feed dealer has your business card or sales letter on the store bulletin board and a supply of them on the counter, right next to the cash register. A good heading might be "Local Breeder Happy to Answer Your Questions about Rabbits."

Talk the workshop over with the dealer. Did the event go well for everyone concerned? How could it be improved? Can you schedule another event later, when you rebound from selling out at this one? It may take a few workshops for both of you to get really good at it, but both you and the dealer will be on your way to increased sales.

When the Phone Rings

When you receive a telephone call as a result of your local advertising, tell the caller the name of the breed or breeds you have. Explain that they are very nice rabbits, pedigreed, kept under the best of conditions, and the offspring of champions (if such is the case). Say that they are reasonably priced but avoid offering quotes over the phone. Tell the buyer that he or she should really come and look them over, and that you would be delighted to show them at the earliest convenient opportunity — perhaps even next weekend.

Letters of Inquiry and Selling by Mail

Your national advertising will bring more letters and e-mail than phone calls. You will find it helpful to answer these with a form letter of information that describes what you have to offer. You'll find the one I use on the following page.

SAMPLE FORM LETTER

Bob Bennett
133 Governors Lane
Shelburne, VT 05482
rabbitalogue@hotmail.com

Dear [insert name]:

I breed and exhibit pedigreed Tans in all four colors: Chocolate, Lilac, Black, and Blue.

These are from the best bloodlines in the United States and England. Years ago when I started with Tans, I spared no expense in obtaining foundation stock. I keep my breeders American Rabbit Breeders Association (ARBA) registered. Several are grand champions. They produce top winners in national shows every year. Ask anyone who has Tans about mine. They know about them. That's one reason why they are My Famous Tans.

All the stock I sell is guaranteed to satisfy or your purchase price will be refunded and shipping arrangements made for return of the animals. I supply full, detailed pedigree papers.

My Tans are kept in all-wire hutches with hopper feeders and drinker-valve-dispensed water. I utilize the most modern management tools and techniques. I stress nutrition, sanitation, and preventive medicine.

Unless other arrangements are made, I ship by air, shipping charges collect. I can estimate charges in advance if you desire.

I drive the rabbits myself to specific flights at Burlington (Vermont) International Airport. In most cases you can pick up your stock at your airport only hours after it leaves my hutches — even if you live in California. The rabbits are in transit less time than they spend in transit to and from the average show.

Should you desire to place an order, send a money order or check, along with your telephone number and the name of the airport where stock should be sent. I will notify you when to expect your Tans.

I have found it advantageous to enclose this form letter because of the volume of my mail. But be assured that I am ready to answer any questions you may have.

Thank you very much for your interest in My Famous Tans. I hope to be able to serve you. My wish is that you will obtain the same enjoyment from my rabbits that I do.

If you run an ad in *Domestic Rabbits* magazine, which is the best medium for reaching rabbit raisers in the United States, its possessions, Canada, and even some foreign countries, you will want to have a good supply of copies of these letters, (and a saved letter on your computer that you can easily attach to e-mails), because your ad will bring many inquiries your way.

When I receive an inquiry in the mail, I reply with the letter and a stamped, self-addressed envelope that the prospective customer can use to reply. I find that this system brings more sales than any other. Computer correspondence, of course, requires no postage.

Additional Correspondence

You may find yourself answering lots of questions about your rabbits, and it may take you several letters before a prospect decides to buy, so allow plenty of stamps and time for correspondence. That's all part of selling breeding stock by mail. Communication via e-mail may speed things up and be more economical.

To ship by air when you receive an order, check first with the airfreight offices for each airline at your airport for crating regulations, as these vary. But you can easily ship rabbits by scheduled airline to anywhere in the world. Depending upon airfreight regulations, you may be able to ship "charges collect" to the purchaser.

When you ship by air to a distant purchaser, you are the one who is picking out the rabbits, not the buyer. That is the way it always should be, because you know better which two rabbits will make the best pair. So it is no disadvantage to the buyer to be unable to visit the rabbitry. You might also prefer the mail-order aspect of the rabbit business, because it lets you conduct it on your own time, at your convenience. Such is not always the case when you entertain visitors at your rabbitry.

You may also sell rabbits at shows, but your best bet for shows will be delivery, not sales. In other words, make the sale by correspondence, and deliver ordered rabbits to the customer at the show, out of convenience to yourself and the buyer and to save delivery charges. If you are a winner at shows, you will receive lots of inquiries. And you may wish to take to the shows pairs, trios, or quartets, or even more rabbits that you intend to sell. Make sure that you select those rabbits you've already chosen to bring to shows for sale; don't get into the habit of deciding to sell rabbits because they're winners, which often is simply because a particular judge liked a particular rabbit that day.

It may pay you to purchase ads in the show catalogs, stating that you will deliver to the show. Take orders by mail or over the phone.

The Ironclad Guarantee

You want satisfied customers. And the best way to find customers who will return is to sell them rabbits that won't. In other words you must sell good rabbits to get repeat orders. Don't sell anybody a rabbit you wouldn't like to have yourself. Of course, that is easier said than done sometimes. Once in a while you will find a customer who is unhappy. It happens to everyone who sells anything.

I find that the best guarantee is one of my best sales tools. If you sell a meat, laboratory, or breeding rabbit, guarantee absolute satisfaction or money back or another rabbit, at the option of the buyer. Meat and laboratory business is built upon repeat orders, and such an ironclad guarantee for these sales is mandatory. It is just as necessary in breeding-stock sales. Even if purchasers don't buy any more rabbits from you in the future, you want their goodwill and good recommendations to others. So make the guarantee I make — absolute satisfaction or else money back or another rabbit, whichever makes the customer happy. That means no quibbling. If the customer doesn't like a rabbit for any reason, even a stupid one, do the right thing: money back or another rabbit.

How Long Should the Guarantee Last?

Fine, you say, but suppose a rabbit dies of old age and the customer wants another one 10 years later. Well, that's ridiculous. And of course, if a customer doesn't take care of a rabbit and it dies a few months later, certainly you have no responsibility. But I have replaced rabbits that died even a year after I sold them, simply to make a friend. That is good business. If I have one hundred rabbits and someone who bought four loses one, it doesn't hurt me a bit to give that customer another. In fact, it does a world of good. The customer often buys more and becomes not just a friend, but one of my best salespersons, telling others that I treat people right.

Let's Set a Price

Could you make money selling dressed rabbits? You could, but of course you have to find the customer.

It turns out that the price of prime or choice steak is being asked and received in some stores. I've seen it in the showcase next to exotic meats such as ostrich, beefalo (a beef-buffalo cross), organic or free-range chicken, quail, guinea hens, and more.

There is a market for these meats because there are people willing to pay the asking prices. Admittedly, the major demand is in the cosmopolitan cities, where some people spend a lot of time thinking about eating something different. Another good place is college and university towns, where many people seem to be adventurous eaters.

My point here is that if you want to sell meat rabbits, you can either let a broker/processor pay you the market rate, or you can more than double your take by dressing the rabbits out yourself and establishing your own list of regular customers who are willing to pay the price that you set. Even if you don't want to dress them out yourself, you could hire someone to do it for you and still come out way ahead of selling live to the processor.

Some rabbit raisers are doing it. They are setting their own prices and finding customers. There are potential customers out there who want something different for dinner, who are already willing to spend more money per pound for other meats that don't even measure up to rabbit in taste and nutritional value. You can find these customers too. Remember, rabbit is the heart-healthiest meat available.

Good Records Make Future Sales

When you sell a rabbit, keep a record of the name and address of the purchaser, and keep a copy of the pedigree. Whenever you get overstocked or have some rabbits you think a past customer might like, you can and should send a postcard or an e-mail. In fact, a good way to make sales is to have sales-pitch postcards made up and send them off regularly to all

---

How Some Rabbit Raisers Set Prices

I KEEP ASKING RABBIT RAISERS what they think about when they sell their rabbits. Here's what some of them have told me:

- "I sell cheaper to 4-H youth. First-time buyers get a free year's member-ship in the breed-specialty club and a 10-pound bag of feed."
- "I will not sell nonshowable stock to 4-H youth."
- "No deformed or diseased stock will be sold at all for any reason."
- "My prices depend on the quality of the rabbit as well as age of the rabbit. I sell rabbits to youth at a better price than to adults. Usually about half price, with pedigree papers."
- "Breeding-quality rabbits are $10 to $15; show rabbits are $25 to $50."
- "Prices are based on $4 per month of age."
- "I sell mostly show stock; the rest go for pets for $15 without pedi-grees. That's why I sell no meat or lab animals."
- "Prices are due to line and to color. Agoutis, Steels, and Blacks [these are French Lops] are mostly $25 to $35. Blues, Tortoiseshells, Fawns, Siamese Sable Points, and Chocolates are more. I have worked hard to get nice, firm-bodied types, and I'm always looking for better and better. I have grand champions to prove it. But better is better. I never sacrifice body for color."
- "I charge $20 for rabbits aged ten to fourteen weeks; $25 for four to five months; $30 for five to six months; $40 to $50 for six to eight months; and $40 to $60 for seniors."
- "Angoras are slightly more than Silver Foxes because of the overhead required to raise those rare colors."

---

your customers and to those who have inquired about your rabbits in the past. Another way to make sales through the mail is to look up names of new members of the ARBA, listed in its *Yearbook*, who live within driv-ing distance and invite them to your rabbitry. I always look up the new members and send them a card. Often, I start them with Tans or Florida Whites. I can give you lots of reasons for joining the ARBA, but this ought to sell you on the idea if nothing else does.

What's the best way to sell rabbits, regardless of their end use of meat, laboratory, or breeding stock? Have them for sale. That may sound silly, because if you don't have any, you can't sell them. If you do, consistently, and have the right kind, you will become known as a steady source of supply. And that's what will make most of your sales for you.

CIVET DE LAPIN, $29

That's braised Vermont farm-raised rabbit with a red-wine sauce served with homemade fettucini.

Phil Brown operates Vermont Rabbitry in Glover, Vermont. He is a processor who buys from Vermont raisers, then sells dressed rabbit to stores and restaurants in Vermont. One supermarket and one restaurant are in my town of Shelburne. The restaurant is Café Shelburne, across the road from the renowned Shelburne Museum. It's a fine French restaurant, and it's the best restaurant of any kind around here.

I go there on occasion. My bride and I have had dinner there on our anniversary for the past 31 years, as long as we have lived in Shelburne. You can't get out of the restaurant for much less than $100 for two. Entrées are in the $20 range, and the menu is à la carte, so you pay extra for soups, salads, appetizers, and some desserts. Of course, if you sample some of the fine wines, you can boost the tab well over a C-note with no trouble.

So Alice ordered the less expensive free-range chicken, even though I told her it scratched in the dirt and manure and ate bugs and worms. I ordered the $29 rabbit, of course. In fairness, both were excellent, but the rabbit was one of the best I've ever had (and I've had a few).

We overheard a couple of other customers order the rabbit, too, and the waiter told me they served seven to eight rabbit entrées just in the time we were there. Phil Brown delivers a fine product that sells well in the face of $25-range competition from lobster, filet mignon, duck, and lamb.

A Growing Wool Market?

ROBERT AND TREVA SAVAGE of Saskatchewan, Canada, raise Giant Angoras, which they like because of the body type and wool production, which is about 2 to 2½ pounds (0.9 to 1.1 kg) per rabbit per year.

"We have a contract with a Swiss company," explains Treva. "They buy all the wool we produce. We are working toward raising five hundred purebreds and earning all our income from wool and stock sales." Treva also tells me the Giant Angora is "beautiful and gaining popularity." Angora wool is a growing area, of course, and perhaps more Giant Angoras will mean more wool. All Angora breeders would do well to investigate them further.

What Else to Sell?

You can sell meat, laboratory, breeding stock, and even pets, if you must. You can sell pelts, and if you have Angoras, you can sell wool (see the box above). But there are other salable items you may not have thought of.

Equipment

If you build all-wire hutches, like those described in this book, you already know they are simple to construct and that you can build one in very little time after you have built a few. Many rabbit raisers build and sell these hutches and derive as much income this way as they do from their rabbits. I have been among them. I have built hundreds of these hutches, offering them for sale with breeding rabbits. I have also put the hutches into pet stores on consignment and sold quite a few that way. I'm not crazy about selling pet rabbits, but inasmuch as thousands are sold, I feel they should have good housing, and so I have been glad to be able to offer these hutches to the owners of pet rabbits, which have become extremely popular.

You may also purchase large quantities of prefabricated hutches and other equipment, such as watering and feeding supplies and nest boxes, at a discount and sell them at a unit price that allows you a profit and yet is attractive to the local purchaser. In some areas rabbitry equipment is not readily available locally, and there is a good opportunity to carry such a line. You can also buy in quantity the wire, hardware, and tools needed to build hutches and sell them to people who want to build their own.

Manure for the Rose Gardener

Rabbits can return so much to you in so many ways. The local rose gardeners flock to a nearby garden-supply center that stocks manure from my rabbitry. The price is preposterous, but the demand is brisk. As more and more people become interested in gardening, and as the supply of other animal manures becomes smaller, the demand for rabbit manure increases. Rabbit manure, even fresh, does not burn plants as other fresh manures can. In the next chapter I'll explain how I use it in my own garden for the flowers and the shrubbery and even the lawn. But for the local garden center, you may want to do as I do: bag the manure in empty feed sacks, and let the rose gardeners and others pay dearly for it. I have calculated that every two bags of manure I sell wholesale buy one bag of feed. And my manure would pay for a lot more of my feed bill if I didn't want to use so much of it myself.

Worms for Fishing

Under the manure that lies beneath my wire hutches wriggles a nice pocketful of spending money. Fish worms sell by the dozen in these parts, and it is no trouble for some rabbit raisers to unload a hundred dozen a week during the spring and summer to a couple of sporting-goods stores. All you need to do is use a pitchfork to turn over the manure and pick up the worms. People who fish don't necessarily want to dig worms themselves, and there are fewer worms to be found in many backyards because chemical fertilizers have chased a lot of them away.

There are several books available on worm culture, and I've read a couple of them. I suppose I could produce more worms if I followed the authors' tips. But I find that beneath the hutches and in my compost heaps are all the worms anybody could want, without my having to make any additional effort. Worms also can help you with your rabbits. In chapter 8 I'll explain how.

Some Sales Advice from Rabbit Raisers

I keep asking rabbit raisers what they sell and how they sell it. It's no surprise to me that they have plenty of good ideas on the subject. Here are some:

John Jennings from Aukum, California, raised New Zealand Reds for more than 50 years. He wrote me just a week before his 75th birthday. Here's how the breeder with some of the best Reds you could buy operated:

> I have to sell at least $250 worth in an order, or it doesn't pay to go to the airport, as it is 160 miles round trip and I have to get up at 4:00 a.m. to make it in time for shipping out at 6:00 a.m. If more than one party wants to order at the same time, that's okay.
>
> Youngsters up to 3 months are $50 each in orders of six or more at a time. A trio of top juniors is $250. Three-hole shipping crates are $25 extra. These rabbits are guaranteed in every way, shape, and form. If they don't prove satisfactory, they are replaced, free of charge. All the customer pays for is the crate and shipping. Very seldom have I ever had to replace a rabbit. They are red, white, and blue pedigreed. And if I say so, I don't believe on the whole there are any better in the world today. They have won more best-of-breed prizes at national conventions than those of any other breeder since I started showing nationally years ago. I usually have around one hundred on hand of different ages at all times.

Ruth Moerschell from Milton, New York, and son Gerry raise Mini Lops, Holland Lops, Rex, and Netherland Dwarfs in a 25-hutch rabbitry.

> One handy way to advertise is to use all those bulletin boards in supermarkets, stores, and laundromats. I carry around a supply of business cards and tacks to post them. I also get a good response by posting a picture of my junior Minis at their cutest (6 to 7 weeks) along with a sheet of tear-offs of my phone number.
>
> One thing I've discovered over the past year is that having cages for sale also really helps sell those bunnies. I didn't want to get involved in cage building. I tried referring customers to a friend who builds cages

and even offered "dollars-off" coupons, but no takers. I helped customers build cages, but some others weren't even interested in that. A friend who had a very successful year selling rabbits sells cages, too, that she builds. She feels this is a big selling point. And in this day and age of one-stop shopping, it appears she's right.

I recall that one of the first questions people have asked is, "What about a cage?" or even "Is the cage included?" So I've ordered my wire and am launching a cage-building business.

Sue Schier lives in California and majored in art in college, which gave her a good background as a hand spinner and weaver. That's why she raises English and French Angoras. Her marketing and barter systems also help put a few of the best on the show table and a few of the culls on the dinner table.

I promote Angoras and spinning at the shows. I take my spinning wheel to the shows. I give free spinning lessons to persons who purchase an Angora from anyone, and the time is well spent because they have to come to my house for the lesson. Naturally, we make a trip out back to look at my show stock and any new bunnies, and I usually end up selling a rabbit.

My brother-in-law has taught me how to neuter bucks — he's a vet. It is illegal to charge money for neutering if you are not a vet, but there is no law against barter. So far I have neutered bucks in exchange for steam cleaning all the carpets in my house, plus haircuts for myself. Neutered bucks make calm, mellow woolers!

Trends in the Rabbit Business

I surveyed about two hundred rabbit raisers and asked them for a breakdown on their rabbit sales; 75 percent sold breeding stock, 46 percent sold meat, 41 percent sold pets, and only 1 percent sold lab rabbits. Of those who sold breeding stock, the average number of rabbits sold in a year was 99, although reports ranged from just a trio to 640. Half of those surveyed sold 50 or more breeders.

R A I S E R ' S   E D G E

More ideas from more of my good friends

- "I'm starting to sell pet-rabbit supplies such as cages and water bottles."
- "We plan to develop a local market for pelts, which will be used for fur hats, mitt liners, jackets, jacket liners, and parka-hood liners. Rex fur is especially attractive for these items."
- "We live in a very sparse (rabbit-wise) area. People here don't really want to pay over $5 for anything, so we sell most of our rabbits at shows. The nearest show is a three-hour drive."
- "Rabbits are a hobby with me. I work as a butcher with retail sales. So I have my own outlet for meat. Otherwise I'd never turn a profit on rabbits."
- "Recently, I contacted a local TV station regarding National Domestic Rabbit Week. I told them I thought a story on rabbit raising would be interesting. It worked! They called me, came out to my rabbitry, interviewed me and a friend, and we were on the six o'clock news with our rabbits. The best part of it was it was all free. It was great advertising."
- "I have cages, feeders, waterers, and extra feed to sell at all times. It adds to rabbit sales."
- "A lot of our sales are local, and we live in a rural area that is very poor. If I lived elsewhere, I could ask more, but here you just can't. I don't usually sell for meat. A large number of breeders go to an animal auction house."
- "Just consider recycling peat moss removed from the dropping pans as peat moss fortified with rabbit manure. Must first find a way to dry the mixture before selling."
- "There was no place for raisers to get supplies in my area, so I opened a rabbit-supply shop in my home, and I am starting to see success from this."
- "I sell very few rabbits for meat. Most of my culls are traded for everything from labor to wheelbarrow handles to nuts and bolts."
- "If you want to sell breeding stock, make sure your agricultural extension office knows about your rabbitry. Send a letter and a few business cards, stating you are available to help people get started."
- "I run an ad in a county sporting newspaper. It goes to every club in the county. One of the clubs ordered 100 pounds of rabbit meat for its 'game' supper. The clubs have suppers in January, February, and March."

The average annual number of meat rabbits sold was 146. Individual reports ranged from half a dozen to one thousand. Selling several hundred meat rabbits a year was common.

Of those who sold pets, the average number sold in a year was 57, about one-third of the meat-sales average. Laboratory rabbit prices were good, ranging from $50 to $100 per rabbit, but there were few sales. Angora wool brought $5 per ounce, and spun yarn commanded twice that. Pelts sold for about $1.

Eighty-one percent of respondents reported some sales from their rabbitry, while the other 19 percent said they were still building their herds but anticipated sales soon. Twenty-seven percent sold breeders, meat rabbits, and pets; a few also sold lab rabbits and had some other categories of sales, including wool, pelts, and, in a growing number of cases, sales to the reptile market.

Nineteen percent sold only breeders and pets. Thirteen percent sold only breeders and meat. Eight percent sold only meat rabbits, and 6 percent sold only pets. Four percent sold only breeders, while another 4 percent sold meat and pets. A few made sales of breeders and pets and kept culls for their own meat supply.

RAISER'S EDGE

Selling Isn't Everything

GEORGE LARUE, a resident of Bluefield, West Virginia, was asked to lend a rabbit to a teacher of young children with learning disabilities.

"I agreed and sent two of my gentlest rabbits, a Blue Rex doe and a Black Tan doe. When the teacher brought them back, she said that my rabbits had really made those kids' day. Some had never seen a rabbit, let alone petted one. It made me feel good to know that my rabbits really helped to brighten these kids' day. I might also add that some of the teachers asked about buying some rabbits.

"I feel that if my animals can bring a little joy into someone's life, then raising them is well worthwhile. Making sales isn't always the main factor in raising my animals (although it's nice)."

What else did these rabbit raisers sell? Many sold cages, equipment, and small sacks of feed that they bought in bulk from feed stores at feed-store prices and resold at pet-store prices, which are much higher. The convenience of having these items available at the point of sale of the rabbit merited the higher prices and also greased the sale of the rabbit, because the sellers provided a complete solution, not a problem of "Where am I going to put this rabbit?"

Other sales included stud and tattooing services, fur pillows, hand-knit Angora garments, ceramic bunnies, trophies, manure, and worms. Some also reported bartering rabbits for babysitting, yard work, cages, equipment, hay, and even guinea pigs.

Twenty-five percent said they were in the rabbit "business." The other 75 percent called it a hobby. Some even thought it was both, which must make tax time fun.

The respondents spent an average of $942 on rabbit equipment and supplies, not counting feed, which makes rabbit raisers a significant market themselves.

Reviewing Some of the Best Ideas

There are many ways to improve your productivity and profit margin. This list sums up some of the best I've found:

- Get rid of mongrels and nonproducers.
- Concentrate on breeding-stock sales. Don't plan to make it big on meat and laboratory sales from a small rabbitry.
- Advertise and ship nationally. There may not be enough people in your vicinity who will appreciate what you've got.
- Find local customers at the feed store.
- Stick to one or two breeds, and become an expert on them.
- Join the ARBA and local clubs for contacts.
- Price your rabbits right.
- Provide books and equipment as well as rabbits.
- Line up sales at rabbit shows ahead of time by writing to all nearby breeders of your breed to let them know what you have available.

- Produce a good-looking sales sheet or folder. Hand it out at shows and at the feed store.
- Put rabbits up for bid as a donation to public radio and television fund-raising auctions. Great publicity. Prime time.
- Advertise in the local all-classified periodicals and "pennysavers." Stay out of large dailies.

R A I S E R ' S   E D G E

Snake Snacks

INCREASING NUMBERS OF RABBIT RAISERS are supplying the reptile market — rabbits for snake food. The notion repels some (of course, so does supplying rabbits for people food), but more and more raisers are embracing this market.

One early supplier of rabbits for snakes got $1 per pound from pet stores, with a minimum size of 2½ pounds. Others who have written to me report receiving prices of $2, $5, and even $10 per rabbit.

"The majority of my reptile customers," a woman who supplies rabbits for snakes tells me, "kill the rabbit just before feeding it to the snake, which I recommend. A large snake that is raised on dead meat is a much safer snake to have around children, and generally is easier to handle. It is also more humane for the rabbit."

She says that live rabbits should never be fed to a snake that is shedding. "The snake will attack the rabbit without attempting to eat it. A snake is most vulnerable when it is shedding and will attack for its own protection. Snakes do not eat when they are shedding." She also says repeat business is the norm. "Snake-food customers are regular year-round repeat customers — a wonderful boost for the rabbit business. Sell a bunny for a pet, and you may never see the customer again, but snake customers return almost every week."

Pat Lamar of the Finley Fuzzy Farm Rabbit Ranch in Washington State explored the reptile market in her newsletter. Here's what she had to say:

"This is a viable market that few breeders take advantage of — including Finley Fuzzy Farm! Unfortunately, the reptile market requires

- Enter rabbits at fairs, and pay the superintendents a commission on sales they make for you.
- Keep names and addresses of breeding-stock buyers. Go for repeat sales by knowing what else they need.
- If you can't sell, swap.

---

selling live rabbits to reptile owners to be fed live to their pet snakes or reptiles, and there is a demand for all sizes of rabbits, depending upon the size of the reptile or snake.

"Although we understand the importance of the feeding of live animals to maintain the reptile's muscle tone, we cannot morally allow ourselves to subject a live rabbit to this end, and thus we are not able to partake of this unique market.

"However, due to this same attitude resulting in difficulty for the reptile owner, there is a market for frozen 'pinkies' and 'fuzzies.' Pinkies are generally newborn mice but can also be newborn rabbits that were born dead or died of hypothermia. Some breeders of show rabbits will actually cull at birth, freezing the unwanted kits. As soon as dead ones are found, they must immediately be put into a plastic bag and frozen. Pet stores that cater to reptile owners will buy them. The reptile owner needs only to make a small cut on the forehead of the pinkie or fuzzy to allow it to bleed, thus enticing the reptile to eat.

"Fuzzies are older baby bunnies that have a complete coat of fur. It is important, however, that none of the pinkies or fuzzies died from disease. Often a doe will squash a baby accidentally, or it got out of its cage and a cat got to it, or it may have died from heat or cold. These are acceptable to the reptile owner, and the rabbit breeder is still able to realize a small margin of profit instead of a total loss on these unfortunate bunnies."

Prices paid for the pinkies and fuzzies are in the $3 to $5 range. Personally, I'm rattled by the whole idea of feeding live rabbits to snakes, and like Pat, I have recoiled from this market.

8

Rabbits and the Home Garden

RABBITS CAN HELP YOUR GARDEN GROW, and your garden can do the same for your rabbits. The manure is great for growing most anything, and some garden produce can help you feed your rabbits. Rabbit manure contains higher proportions of nitrogen and phosphorus than many other manures and more potash than most. It won't burn plants or lawns even when applied fresh. It comes in a convenient round, dry form. I can't get enough of the stuff, and everybody keeps trying to talk me out of it. Every spring my gardening neighbors expect at least a feed sack full (along with the tomato plants I give them).

Make Manure Go a Long Way: Compost!

You can use rabbit manure just the way it is, forked up from under the hutches. But composting it with other materials will make it go farther and enhance its value as an all-around fertilizer and soil conditioner.

Twice a year, I push garden cartloads of manure to an area near my garden and build compost heaps. In November, when all the leaves fall from my large trees, I build piles of alternating layers of rabbit manure and leaves, along with plants from my fall garden cleanup, with interspersed sprinklings of superphosphate, lime, and sometimes chemical fertilizer. In spring I build additional heaps, using more manure and leftover leaves plus those from the spring cleanup. Later I add grass clippings and weeds to the heaps, and thus I have a supply of late-summer compost to add to my spring output.

Follow any good composting scheme, add rabbit manure, and watch things grow. If the compost doesn't rot fast enough for me, I get out my shredder and grind it all up. My favorite composting book, which has complete instructions, is *Let It Rot!* by my late friend Stu Campbell (Storey Publishing, 1998).

In 1978 I started vegetable and perennial flower gardens in the worst clay soil I could imagine. Some of it actually was not just brown but purple and green. I probably could have mined it and sold it to a hobby store for clay modeling. It was a terrible garden situation. All the vegetables had to be grown on raised beds. Most of the first perennials died from lack of drainage and from being forced out of the ground by frost heaves. When this stuff was dry, the garden was like a brickyard. When it was wet, you'd be afraid you would sink in and not come back up. But I have put rabbit manure on it every spring and every fall since then, and continue to do so, because in only a few years amazing improvement took place.

Now my vegetable garden soil is soft and crumbly, rich and black. The drainage has improved, and the yield is wonderful. It's a cinch to till because of a texture that would remind you of chocolate cake, and it absorbs water like a sponge. If I had to move, the worst part would be leaving this soil. In fact, one rabbit raiser I know, who achieved garden soil like mine through years of application of rabbit manure, hired a backhoe and a dump truck and took his vegetable-garden soil with him when he moved to another part of town.

Where and How to Use Manure

Straight rabbit manure or manure compost can be dug and tilled into any garden soil with good results — not just clay soil like mine but also sandy or any other kind of soil. It improves the texture of the soil, making it more receptive to rain and generally improving the flowers or vegetables. I add lime to the leaf-manure compost for most uses, but around the azaleas and rhododendrons, which like an acid-soil environment, I leave it out. Usually one of my compost heaps is an acid one, while the other is sweetened with lime.

If you dig the compost into your flower beds before planting time, you'll soon have a good stand of whatever flowers you want to grow. Mulch your plants with more compost. For my flower beds, I sift the compost to a fine consistency and mulch the beds quite neatly. The mulch conserves moisture, smothers weeds, and keeps the beds attractive and nearly maintenance free. The following spring I dig in this mulch with some additional compost and start all over. I don't see how you can put in too much rabbit-manure compost.

Into the vegetable garden goes cartload after cartload of compost before rototilling in the spring. Into each planting row goes some sifted, shredded compost. Into each hill of vine crops, such as cucumbers, squash, pumpkins, and melons, goes a bushel basket full of compost. Vine crops really thrive on the stuff. When everything is up and growing, I don't have to hoe weeds. I smother them and conserve moisture by banking a compost mulch around every row and hill. The next spring, just as in the flower garden, this mulch is tilled under, and the whole process starts again.

Rabbit Manure Saves Energy, Too!

In days of yore, before the advent of heating cables, gardening hotbeds often were heated with manure. A few years ago I found an old gardening book with directions and substituted rabbit manure for the horse variety. Now I start my tomato and pepper plants and my impatiens in late March here in Vermont in a rabbit-manure hotbed made of old storm windows, into which goes about a 2-foot (0.6 m) depth of droppings. By Memorial Day, when the danger of frost is past, I have hundreds of vegetable and

---

KEEP IT COMPOSTED AND DRY

Rabbit manure will not create any odor problems for you if you keep it dry and composted. If it sits in a pit, soaked with urine, you'll have trouble handling it. Well-drained, graveled beds under the hutches will keep it dry.

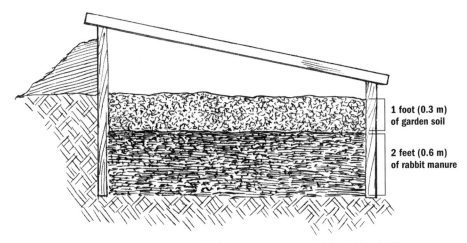

1 foot (0.3 m)
of garden soil

2 feet (0.6 m)
of rabbit manure

A rabbit manure hotbed is just a cold frame over a deep pit filled with rabbit manure.

flower plants. Electricity bill: $0. The true test of the capability of rabbit manure to heat a hotbed when the temperature is way below freezing outside is the impatiens. This annual flower seed, which I save from year to year for plenty of color in the shade, takes about 3 weeks to germinate. The sustaining bottom heat of the rabbit manure does the job.

Help from the Worms

Fat, wriggly red worms inhabit the compost heap, partly because they are attracted to it and find it a comfortable and palatable environment, and partly because I put them there with the manure when I clean out my rabbit barn. These worms do a great job of converting the compost into good, rich, black potting soil. For starting seeds I know nothing better. Down at the bottom of the compost heap you'll find good potting soil just for the digging.

What about the Lawns?

Rabbit manure and rabbit compost can also go on the lawn as a top dressing at any time of the year, but a little caution should be taken. If you feed pellets or recleaned grains, the manure will not contain weed seeds. But if you add hay and straw to your rabbit feed, you might not like the looks of your lawn if you use rabbit manure.

SOME GIANT TOMATOES

John W. Turner of Decatur, Georgia, does not raise rabbits, but rabbits help him raise some of the biggest and best tomatoes in the United States.

He does a lot of things to grow them big. How big? His plants are 6 feet apart, and his support trellis for each plant is 8 feet high. At the Georgia State Fair, he's won Heaviest Tomato awards over and over again.

John uses rabbit manure along with such fertilizers as Tomatoes Alive, Nature Meal for Tomatoes, Tomato Mate, granite meal, Epsom salts, dolomitic lime, and even Kricket Krap (Bricker's Organic Farms, Inc., Augusta, Georgia) — a great name for the excrement of crickets raised as fish bait.

One final note on names for manure: A proud rabbit raiser calls the excrement from rabbits "bunny berries."

I get some weeds with the rabbit manure, but I don't mind them much. Years ago, when I had a suburban lawn and used plenty of commercial fertilizers and weed killers, I still had some weeds. Now I have a much thicker, healthier turf and probably no more weeds than I had years ago. If I really wanted to knock them off, I could use a spray weed killer. One of these days I might even do it.

For the lawn I either run the compost through my shredder to make it fine enough for my fertilizer spreader to handle or I sift it through a frame of 2 × 4s and hardware cloth. You don't want it so coarse that it will mat and smother the lawn. The compost for a lawn must be fine enough to let the blades of grass through. I often simply "broadcast" dry manure with a shovel from my cart. A friend of mine spreads dried rabbit manure straight on his yard all winter long. He's the first one in town to crank up the lawn mower in the spring.

What the Garden Can Do for Rabbits

I insisted in chapter 5 that you feed rabbit pellets. And I stand by those statements. But the garden can supplement the pellets by providing roots, greens, grain, and seed crops that you'll find useful as you progress in rabbit raising. Some of them are particularly useful for kindling does and show stock. You know already that I don't suggest anyone try to feed rabbits exclusively on the output of a garden. But since my rabbits put so much into my garden, I figure they deserve to get something out of it.

Sunflower seeds make an excellent fur conditioner. All rabbits love them, but they are especially good for those you plan to show. Sunflowers are a cinch to grow at the back of the garden, perhaps among the corn, where they won't shade lower-growing vegetables.

Corn is great for rabbits. Corn is also great for humans, so perhaps there won't be any left for the stock. But rabbits like the dried corn stalks and shucks and, if you can spare them, a few ears. After they have dried, they will soon devour them, cob and all. A few green shucks won't hurt adult rabbits either, provided you don't give them too many.

Pumpkin seeds are eaten by some rabbits, snubbed by others. We always grow some big jack-o'-lantern pumpkins, and we dry the seeds for the rabbits.

Mangel beets have been grown for years and years in Europe as a livestock feed. These are huge beets that I leave in the ground and cover with leaves into late fall and winter. Then I slice them for the rabbits. A number of years ago, I had a nicely marked Dutch buck that never seemed to put on enough weight to win in a show. He did as poorly as 11th in a class of 11. I lent him to a friend one fall, and by the time I got him back in the spring, he was so plump that he won best of breed out of 77 Dutch rabbits. Mangels had done the trick over the winter. Since then, I have fed these beets to rabbits who need to put on a little weight and don't seem to do it otherwise. You may have to check the seed catalogs closely to find mangel seeds, but they are available.

Rutabagas can be grown with the mangels and left in the ground, too, covered with leaves into late fall and winter. Whenever I need them, I just dig them out. I feed these to adult rabbits that like them better than mangels. Rabbits are individuals and have individual tastes. Nobody likes turnips in my family except the rabbits and me.

Carrots are a rabbit favorite. If rabbits will eat anything, they'll eat carrots. The problem is that humans want them too. I usually plant lots more than the family wants and feed them to newly kindled does and my favorite old bucks. I dry the tops like hay and feed them to one and all. When I thin the young carrots, I dry the whole business and also feed them to all the rabbits.

Lettuce thinnings are also dried for all or fed green to adults very sparingly. A new mother will always get some lettuce.

I have dried bean and pea vines like hay and fed them to all.

If you have the room, grow your own alfalfa, clover hay, and even soybeans. Can you grow oats? If so, you've got grain and straw.

Remember that these feeds are merely supplements. If you want to put meat on your rabbits, if you want top flesh condition, you must feed pellets. Grains, greens, and roots are treats in addition to the pellets — don't try to make them do too much.

DON'T FEED GREENS

Remember that young rabbits can't handle green feeds. It may give them diarrhea.

Other Tasty Treats

The home garden isn't the only place these and other feeds are found. You can ask your vegetable store to save you the tops and the castoffs, but be sure to wash them well. Feeds pop up in other places. One friend of mine lives near a huge bakery. When the freight cars roll in to unload the wheat, some spillage occurs. He scoops up about five pounds of whole wheat a week and uses it to supplement the pellet ration.

Dandelions, plantains, dry bread, leftover cereal (with milk!) — these are other feeds that can supplement pellets and are better fed than wasted, particularly to nonproductive animals who may be lazing along through the summer. These feeds won't do much for great production but are helpful for maintenance.

9

Preventing Problems

GOOD ROUTINE MANAGEMENT and preventive measures will help you keep your rabbits productive and healthy, your rabbitry attractive and sweet smelling. Such practice will minimize four basic types of problems in rabbit raising: breeding failures, disease, poor sanitation, and unattractiveness. The first can prevent you from even starting in rabbit production; the second can stop you in your tracks; and the last two can incite a zoning board to put you out of business.

Breeding Tips

There is only one breeding problem in rabbits: It's when they don't. To dramatically reduce the likelihood of that occurring, following are a few tips.

- Keep bucks and does in good flesh and fur condition. Don't let those does get too fat.
- If you work at a regular job Monday through Friday, mate your does on weekends at 6- to 8-hour intervals. You may then expect litters on Tuesdays and Wednesdays. That means you'll have time on the weekends just before to prepare nest boxes. If there are children in the neighborhood, they will be in school when litters are kindled, making for a quiet birthday. Keep a close eye on hutch cards and calendars.

- Always check vulvas for that reddish purple color that means does are ready for service. Always check testicles to make sure they are fat and full, indicating virility.
- Test mate your does a week after first service and learn which does test mate positively. Learn to palpate does. (See chapter 6 for more information.)
- Try not to handle does 3 weeks after mating in case they are pregnant. And don't allow visitors into the rabbitry when does are due to kindle. Keep things nice and quiet at kindling time.
- The day after kindling, distract the doe with a green tidbit before inspecting the litter.
- Use extra insulation and nesting material in cold-weather nest boxes. Provide heat with a light bulb or nest-box heater if it is extremely cold.
- Use a well-ventilated nest box, especially in hot weather. The type made of cage-floor wire works well.
- If you should lose a litter, rebreed the doe within a couple of days. If she kills her litter, give her a second chance. If you want, give her a third chance. But three times and she's out.
- Never mate two rabbits with the same fault: e.g., poor fur or body type.

Let There Be Light

I wish every day could be June 21, the first day of summer and the longest day of the year in the Northern Hemisphere. That's because rabbits breed almost unfailingly around that time of year.

Some say springtime warmth is what makes the difference, but for years researchers have known that really it is day length. Egg producers keep the lights on in the henhouse 16 hours a day all year long. Ideally, rabbit raisers should do the same thing. The idea is to make the hens and rabbits think every day is June 21.

Inside my barn, a timer puts lights on in the late afternoon during the fall months. For years, these were just regular fluorescent and incandescent lights. But because I also like to start vegetable plants in the early part of the year, I learned about full-spectrum lights.

It's true that you can start plants under a fluorescent light that costs very little. So why would you want to pay as much as $10 or more per bulb? That's what I pay now, and my plants absolutely flourish, easily outperforming any I started under the cheap bulbs.

But it's not just plants that react well to full-spectrum light. Humans and rabbits do, too. You may be familiar with seasonal affective disorder, or SAD, which can cause feelings of sadness or depression. If we are depressed during the dark months, it could be because we are not getting enough sunlight. The full-spectrum bulbs simulate the sunlight.

If more light works for chickens, it should work for rabbits, too, and it does. I put some in my rabbit barn, and the conception rate increased. If your rabbits are in a shed or barn, think about putting some full-spectrum lights on in the fall to increase day length, at least in their minds. Many rabbit breeders have difficulty producing litters in the fall and winter. This problem could be solved easily with added lighting, and I strongly recommend the full-spectrum bulbs. You can find them in feed stores, garden centers, and lighting stores, or by mail-order from your favorite seed company.

Disease Prevention

Of course, always start with healthy stock that exhibits a bright eye, glossy coat, firm flesh, and vigor. Isolate all new stock from the herd for a week before introducing it into the rabbitry. A pen in the garage or tool shed or near the garden may be necessary.

Maintain a good feeding schedule of the proper feeds. Never give greens to very young rabbits. How many times have I said that? Have I told you why? Mostly it's because it's dangerous to change the diet of a young animal abruptly. If you are giving it dry pellets, hay, and water on a regular basis, that's what its stomach expects. If instead you give it something wet and gassy, you're apt to wind up with a dead rabbit.

Keep the water fresh and pure. If you have crocks, rinse daily, and wash and disinfect weekly. If you use an automatic system, observe valves daily, and flush the system at least once a month — more frequently if hard water causes a sediment buildup. If rabbits aren't eating, something's wrong. First make sure they are getting enough water — rabbits without water won't eat.

Check for that bright eye and glossy coat, feel for the firm flesh, and observe activity. Check footpads occasionally to avoid sore hocks or ulcerated footpads. If one rabbit is a habitual stomper, put a flat board in the hutch for it to stomp on. Watch the droppings. They should be large, round, and firm. Take steps at the first sign of any diarrhea. Watch for runny noses, and keep an ear out for sneezes. Matted fur on the insides of front paws means rabbits have been wiping runny noses.

Don't be afraid to dispose of diseased stock. If you have rabbits with colds or pneumonia, it may cost you far more to cure them than to replace them. And I don't mean just the cost of medicine or even the veterinarian's fee. They may bequeath their poor health to offspring and generate a whole herd of health problems. In the meantime diseased rabbits can infect others in the rabbitry. Most disease problems — and there really are few that hit rabbits hard — can be prevented. Some can be treated successfully. But some aren't worth fooling around with. A swift blow to the skull is a foolproof cure.

Routine Preventive Medicine

The main thing to keep in mind with herd health is to maintain it, not recapture it. Recovered rabbits often don't ever measure up to those that have always enjoyed perfect health. They don't grow or mature as they should, and they often aren't worth the time it takes to treat. So stop disease problems before they start by following a good routine of *preventive* medicine.

Here's the routine I follow, one that has kept my closely quartered stock in good health for years.

- First I make sure my pellets contain copper sulfate (most do) to help control coccidiosis parasites and prevent diarrhea. Check your feed tag.
- Once a week I treat the water with the original and most water-soluble brands of oxytetracycline — Terramycin or Neo-Terramycin Soluble Powder — for growth promotion, for diarrhea prevention, and because I'm absolutely certain it increases litter size.
- Before each litter emerges from the nest box at 2 to 3 weeks of age, I disinfect and rinse the hutch floor.

This routine may sound like some trouble and expense. But I can assure you that these measures are cheap and trouble-free if you look at the health record my stock has maintained: very little diarrhea; no sign of coccidiosis, a serious condition caused by a debilitating parasite; not a sniffle in years; drastic reduction of mortality of young stock due to mucoid or nonspecific enteritis; practically no mortality of mature stock; and larger average litter size than anyone else who raises my particular breed in the United States and elsewhere in the world, as far as I know, from correspondence with other breeders of Tans.

Let's take a closer look at this preventive medicine — the best medicine.

Broad-Spectrum Antibiotic Treatment

Terramycin and Neo-Terramycin are broad-spectrum antibiotic soluble powders without peer in growth promotion and stress and diarrhea (also called scours) prevention in my herd. Treatment of the water for one day a week has also boosted litter size, which is not surprising because it keeps the rabbits functioning in top condition. It also helps them convert feed better. Terramycin has been in use since its discovery in 1949 — the first of the world's broad-spectrum antibiotics. The water-solubility of Terramycin and Neo-Terramycin is unmatched and is especially important when you use automatic watering systems; they are not likely to clog the system.

I use a teaspoon of the soluble powder in every 5 gallons of water. I also water rabbits with Terramycin or Neo-Terramycin before and after a show, before I ship them, and with the onset of a drastic change in the weather, all to alleviate stress. Again, get these broad-spectrum antibiotics from a feed store. I have been using them faithfully for more than 40 years! I give the treated water to all the rabbits, regardless of age, but I make sure to withhold it for 2 weeks before any rabbits go for meat, to make sure that the antibiotic is eliminated from the system.

Disinfecting

Before the litter pops out of the nest box, I use a Lysol solution, following the directions on the bottle, to scrub the wire-hutch floor, then rinse and rinse and rinse with clean water to minimize the chances that the litter will pick up any disease organisms that may be lurking there. Other disinfectants work as well. Some raisers sear the floor with a propane torch.

Coccidiosis

Rabbits with coccidiosis, a condition caused by infestation of the parasite coccidia, do not convert feed efficiently and fail to gain weight properly. Infection with these parasites also makes rabbits more susceptible to other intestinal problems. Packers of rabbit meat find coccidiosis, or spotted liver, to be quite prevalent. Not only must packers discard the liver, but more to the point, some of them won't buy any more rabbits from raisers whose rabbits have exhibited coccidiosis.

Mucoid Enteritis

Mucoid enteritis is a form of diarrhea that hits young rabbits primarily between 5 and 8 weeks of age. The cause of the disease is unknown. It can kill them in 24 hours. Today the young bunny looks great, bouncing around the hutch. Tonight it's thin, scruffy, wet around the mouth, squinty-eyed, and sitting listlessly in a corner, perhaps with its feet in the water crock. It usually emits a steady stream of jellylike feces that foul its rear underside completely. If you pick it up and shake it up and down, you hear a splashing sound within. Tomorrow morning, it's dead.

And the next day another may go. And another, a couple of days later. You could lose an entire litter.

Preventing Coccidiosis and Mucoid Enteritis

You can do several things to help prevent mucoid enteritis and coccidiosis at the same time. Sulfaquinoxaline sodium was the treatment of choice for mucoid enteritis for at least 60 years, but the Food and Drug Administration has taken it off the market. In recent years feed manufacturers have added copper sulfate to their feed formulas, and they also now include yucca schidera extract in pellets. Look for these ingredients on the feed tag. They reduce the incidence of mucoid enteritis, as well as coccidiosis, which often leads to mucoid enteritis. So does routine feeding with low-protein, high-fiber pellets, along with oats and hay.

The regular use of the antibiotics Terramycin or Neo-Terramycin helps to maintain a strong animal whose health is less likely to be compromised by parasites. Clean, dry, draft-free wire cage quarters, and a nest box that keeps the bunnies inside until as close to 3 weeks of age as possible are also important preventive measures.

While you can take steps to prevent it, there's no sense in trying to cure animals infected with mucoid enteritis — especially not the young ones. Even if they live, which happens occasionally, they never amount to much. Their hindquarters are pinched. Overall, they are unthrifty; their growth is retarded. And you wouldn't want to breed such an animal, as it might pass its susceptibility along.

Dealing with Other Health Problems

For so-called snuffles, which is actually a highly contagious respiratory ailment, and pneumonia, use a higher level of Terramycin or Neo-Terramycin: 1 teaspoon per gallon of water. First, isolate the animal to avoid contagion. Snuffles can go through a rabbitry like wildfire. If the rabbit doesn't respond in a couple of days, give a daily intramuscular injection of penicillin, following instructions for dosage that relate to body weight, until symptoms disappear, and then for 3 days longer after that. The "3 days more" cannot be overstressed: just because symptoms disappear doesn't mean the problem has.

There are topical ointments sold that purport to cure snuffles. In fact, they only mask the symptoms. I would avoid these because they can really cause more harm than good. One of them was a popular chest rub for children even before World War II.

Treat sore hocks as you would any abscess, using tincture of iodine or an antibiotic ointment. Give the rabbit a board to sit on. Sore hocks often may be caused by thinly furred footpads. Such animals are not good sire or dam candidates as they may pass this weakness along. Another candidate for sore hocks is the high-strung rabbit that stamps its feet a lot. Again, I wouldn't want to breed such a characteristic into the herd.

For ear canker, a scabby condition caused by parasitic mites, drop some mineral oil or even salad or cooking oil into the ear daily until the mites that cause it are saturated and the crusty scales disappear. But to prevent ear canker, use all-wire hutches. Because my hutches have wire floors, I've never had a rabbit with ear canker, and you shouldn't, either. Parasites are born in dirt and manure, which is why rabbits raised on solid floors are often afflicted with them.

Ringworm is the only rabbit illness I know that is contagious to humans, and I haven't seen any of that in 30 years. You would have to touch an infected rabbit to contract it. My best advice is to try to prevent disease and other health problems, rather than to treat or try to cure them.

Is Medical Treatment Available for Rabbits?

In the United States the Food and Drug Administration and its Bureau of Veterinary Medicine must approve all medications for animals by species before they may be sold. Few medications are approved for rabbits, because few drug companies are willing to stand the great expense of testing for this species with little hope of return. They do not view the rabbit market as large enough to warrant the expenditure, so they don't test their products for rabbits, and the government doesn't approve them for rabbits. But that doesn't mean the drugs don't work on rabbits. They do, and you can buy them because they are tested on rabbits for other species. The big problem is that the rabbit industry has not sold itself as a market, and therefore the drug companies don't know just what opportunities are available for selling the products. An industry-wide market-research study would solve the problem. It would be a good project for the ARBA to fund.

Few veterinarians saw rabbits before the 1970s, when Merck, a large drug company, asked me — only an English major with just a practical knowledge of rabbits — to write the rabbit section of the Merck Veterinary Manual, a handbook for veterinarians. Since then, pet owners have taken their rabbits to the veterinarian in steadily increasing numbers. Quite a few veterinarians now treat rabbits effectively, and things have improved for rabbits. Internet vet networks have also appeared. The fact remains, however, that veterinarians are not going to tell you how to treat your rabbit, because they want to do that themselves and get paid for it. Few rabbit raisers can afford these expensive veterinarian visits, and it often makes no sense anyway. It is usually better to dispose of the rabbit. If you have a dead one and are mystified and fearful for the rest of the herd, though, it could well be worthwhile to pay the veterinarian to do a postmortem examination.

Ironically, drugs approved by the FDA for humans and other animals usually are tested on rabbits first. But drug companies can't help you with medicines created specifically for your rabbits; the law won't allow it. Therefore, rabbit producers are left to choose between folklore that advocates vinegar and plasterboard and medicinal products approved only for major food-animal species (cattle, hogs, and poultry). Worse, there are a few scam artists out there making claims that only take your money and perpetuate the problem. Always be on the look out for the word "treat." When a product is said to "treat" a problem, I am immediately wary. I want only to prevent or cure problems. Treating them just keeps them going and makes me poor.

Sanitation

Clean, dry hutches in a well-ventilated rabbitry will do more to maintain herd health than anything else. Keeping an all-wire hutch clean and dry is no problem. The urine and manure go right through the bottom. If they land in a well-drained pit below (topped with gravel, sand, or stones), inhabited by worms, two things will happen to help keep the rabbitry sanitary. The droppings will remain dry, in the first place, to hold odor down. And the worms will consume them, turning them to rich, black potting soil, further reducing odor and the incidence of flies.

I don't recommend removal of manure from beneath the hutches daily or even once a week, particularly in warm weather. Moving it around causes unnecessary odor, and it also prevents the worms from doing their job. I remove manure only twice a year, in the fall and in the spring, on cool, calm days. I do recommend a sprinkling of lime or superphosphate on the droppings once a week, especially in hot weather. That will hold odors down and improve the fertilizer value of the manure. Ordinarily, the worms and the lime or superphosphate eliminate the odor problem.

If smells should persist, however, there are a number of products available to help you. I prefer the dry granule forms that you can sprinkle on the manure to the type you mix in water and pour or spray, because I believe in keeping the manure and the whole rabbitry as dry as possible. Dampness, of course, does not help eliminate odor, and neither dampness nor odor is good for animal health either. There is really no excuse for a bad-smelling rabbitry.

What about Flies?

Everyone knows flies breed in manure, among other places. But few flies appear in a rabbitry where the manure is dry and the worms are working. If they do cause a problem, spray the manure beds with a fly spray. Spray down on the manure so as not to contaminate the rabbits or their feed and water. It's a good idea to turn water crocks over and to spray before feeding pellets, when feeders are empty. Spraying is less work if you have a valve watering system because it is closed and protected from contamination.

I used to spray manure beds three or four times a week during the hot summer months, until I added a greenhouse to one side of my barn. In summer, when there are no plants in the greenhouse (they are in the garden then), it is a nice warm place for spiders. The flies zoom right into it, seeking light and warmth. What they get instead is death by spider. I love my spiders. In summer, I call my greenhouse my "web site." I don't have to spray anymore.

Mice

Mice can be a problem if you let them. If feed stands around in sacks, unprotected, mice will soon find it. I keep my feed in covered, galvanized garbage cans. Use of hopper feeders, which prevent spillage to the ground below, doesn't just save feed. It also keeps mice away. They will burrow under manure piles and come up looking for feed if it is spilled regularly. If you are bothered by mice, put your pellets in metal cans for storage, and don't overfeed to the point that feed stands around waiting to be eaten by mice or gets scratched out by the rabbits. Then get a good rat poison and place it around the holes to their burrows under the manure.

Complaints from Neighbors or Relatives

A neat, attractive rabbitry, or one that can't even be seen at all, will minimize or prevent complaints from neighbors or from a spouse or parent who has less enthusiasm for the creatures than you have. I like to keep a low profile in my neighborhood as far as rabbits are concerned. While hundreds of thousands of rabbit raisers have read my books, most of my neighbors don't even know I have rabbits or write books about them. Zoning laws do permit rabbits where I live, but I don't realize any income from my neighbors, so there is no sense in talking up my rabbits with them.

My rabbit shed in New Jersey sat among trees and shrubbery. I kept it painted green to blend with the surroundings. It looked like any garden tool shed in town. It could hardly even be seen from my house in the summer when the trees wore their full foliage. A fence, flowers, or climbing vines would have done the job as well, and I recommend such shading protection and camouflage, especially if you live in a suburb or residential area of a town or city.

If you don't have a building, a good fence is a must to keep out marauding dogs. Much grief has been caused in rabbitries by dogs.

Maintaining good relations with your neighbors is all important. If your neighbors don't like you, they may complain to the authorities about your rabbits if they know about them. Many zoning laws are ambiguous, and if the complaint is strong enough, it won't matter how neat and clean your rabbits are. You may be forced to get rid of them. So when you start with rabbits, don't make a big announcement in the neighborhood. Keep a neat, clean, odor-free rabbitry full of healthy and attractive rabbits, and stay in the good graces of your neighbors.

Other members of your family may not enjoy rabbits as much as you do. But you can win them over in various ways. Income from my rabbitry, for example, has gone toward college expenses for my three children. And in the meantime the rabbits can endear themselves to the entire family by providing extra money for certain luxuries and treats you might not otherwise enjoy. Our rabbits have bought the family restaurant dinners, baseball and Broadway tickets, a grandfather clock, a 10-speed bicycle, and, most recently, Queen Anne furniture. Members of my family have learned to tolerate, if not appreciate, my rabbits.

10

On to the
Rabbit Shows!

SHOWING RABBITS IS ONLY ONE REASON for raising them, but for some raisers it is the only one. They love the fun and thrill of competing and winning. They enjoy spending a day with friends who have the same interest. They often make show day a family outing, with a tailgate picnic, a fishing trip, or local sightseeing.

One of the most rewarding aspects of rabbit shows is learning from the other raisers and the judges what you have done wrong and right so you can overcome your mistakes and oversights before another show date arrives. Raisers who hope to sell breeding stock try to build a record of winning that will attract customers.

What can you win? There are trophies, ribbons, and special prizes put up by other competitors. There are breed-club sweepstakes points and grand championship leg certificates and, simply, the judge's nod. Except for some state fair exhibitions, there is not a lot of money to be won, and that probably does more than anything else to keep rabbit shows as honest and fair as humans can make them. You will always find friendly competition and none of the brutally serious attitudes that prevail at dog or horse shows.

Locating and Entering Shows

Rabbit shows do not receive a lot of publicity in local newspapers. That's probably because they are not spectator oriented; they really are fun only for the exhibitors or those who plan to exhibit.

Local and state clubs and breed clubs sponsor the shows. If you belong to local and breed associations, you will learn about rabbit shows near you via club newsletters. If you belong to the ARBA, you will find a listing of upcoming shows in each issue of *Domestic Rabbits* magazine and on its Web site: www.arba.net. The listings include the show date and the name and address of the show secretary.

Contact the show secretary and ask to be put on the postal or e-mail list for the show catalog. There is no charge for this catalog, which gives directions to the show, the time to arrive, the entry fee, and the prizes offered and lists the breeds that are sanctioned, which means those breeds that will receive sweepstakes points. Any recognized standard breed may be shown, whether listed or not.

Fill out the entry blank and mail or e-mail it as early as possible to ease the workload of the secretary, who is a volunteer. Normally, I don't enclose the entry fee in case some last-minute conflict prevents me from attending. This fee, usually about $3 per rabbit, can be paid upon arrival at the show. Make a copy of the entry blank so that you won't have any mix-ups as to which rabbits you are entering.

On the day of the show, pack up your rabbits in crates or carriers that are large enough to allow them to turn around but small enough to fit in the back or trunk of your car, where they will ride nicely if the day is not too warm. Most rabbit shows take place in the spring and the fall, but on a hot day keep your contestants as cool as possible. I use all-wire carriers that provide the same ventilation my hutches do. They are easy to make or easily purchased.

Most show secretaries specify in the catalog that you must arrive at least an hour before the judging starts. Upon arrival get in line for your show cards, and pay your entry fee if you haven't already.

At some shows, rabbits are cooped and earmarked by club members who have been assigned the task. At the best-run local show I've ever

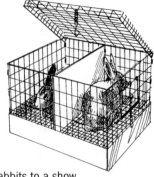

All-wire carriers are great for transporting your rabbits to a show.

attended, Tony Pisanelli's Green Mountain Rabbit Breeders spring show each Memorial Day weekend, rabbits were unloaded, fed, and watered. But at most shows rabbits remain in carrying cages, and it's up to you to bring along a felt-tip pen for marking their right ears with judging numbers and to fasten judging-remark cards to your carriers.

Be alert to announcements by the show superintendent, who will call for the various breeds to be brought to the show tables. It is your responsibility in most instances to carry your own stock to the table.

Judging by Class

Judges evaluate rabbits by classes, grouping them first by breed, then variety, age, and sex, in that order. If you raise New Zealand Whites, for example, the first class on the table will be senior bucks; the next, senior does; then 6- to 8-month bucks; 6- to 8-month does; junior bucks; junior does; and, perhaps, prejunior bucks and does. Seniors are 8 months and over; 6-to-8s are, not surprisingly, 6 to 8 months; juniors are under 6 months; and prejuniors, if any, are usually about 3 months of age.

Showroom classes for smaller breeds, such as Tans, are simpler. Seniors are 6 months and over; juniors are under. In some shows does and litters and meat pens of three fryers each vie for prizes. Fur classes are also special, with separate entry fees. The prize money is a percentage of the entry fee. In fur classes, normal white fur, normal colored, Satin, and Rex furs are judged separately. I like to enter Tans, which go in normal colored fur classes, because invariably they win. Tan fur, although classified as normal, exhibits greater sheen than that of any other rabbit except the

Havana, including the Satin. And you don't see too many Havanas, which were used to make the Tan, entered in shows. It's not unusual for Tans to cop all the top places in normal colored fur. I expect them to be put in a class of their own, like Satins and Rex, one of these days. In the meantime, at the expense of the other breeds, Tans are cashing in.

First in the Class

A class of rabbits might contain, for example, twelve senior bucks. The judge looks them all over carefully and measures each individual's worth by two yardsticks. One is the standard for the breed, which he should know by heart but usually has nearby for occasional reference. The other is comparison — one animal against another. One by one, the rabbits are placed, from 12th to 1st, and the judge dictates remarks about each one's good and bad points to a recording secretary, who writes them on a show card that becomes the property of the animal's owner. The show secretary indicates placings on a show-record sheet, which becomes the basis for awarding trophies, ribbons, and special prizes and for the report to all exhibitors and to the breed clubs. If the show secretary does not issue you a complete report of the show (most report only to you what you have won), you can find a comprehensive account in the newsletter of the breed-specialty club. To receive this newsletter, of course, you must be a club member.

Victory!

Let's say your New Zealand White senior buck wins its class of 12. If there were at least two other exhibitors, you have won a "leg" on a grand championship. Two more legs win you a grand championship certificate. But that means winning at two more shows. It's not easy.

Your first senior buck is held aside by the judge until the first senior doe, 6-to-8 buck and doe, and junior winners are chosen. Then he or she looks all these rabbits over together and chooses the best New Zealand White. If that's your buck, it becomes best of variety, probably wins a rosette ribbon, and is held aside until the class and variety winners of New Zealand Red and Black varieties are chosen. The judge then selects the best New Zealand. If your New Zealand White buck wins best of breed, you'll probably get a trophy. If there is to be a best-in-show winner, your buck can compete against the best of the other breeds, perhaps for a silver cup or a similarly sought-after prize. If you take that home, you doubtless will consider your time and effort at raising rabbits well spent indeed.

Actually, there are other prizes to be won. If, for example, your buck had lost out to a doe for best of variety or best of breed, then it might have won the consolation best of opposite sex, which usually gets a trophy also. Right now, I'm counting four trophies won this spring, as I have attended four shows and won best of breed at two, and best of opposite sex at the other two. I'll attend one more show this spring, the Green Mountain show in Vermont, and I hope to maintain my winning place at the expense of all the fine northeastern United States and southern Quebec breeders.

If I do, I'll be pleased but not too jubilant. And if I don't, I'll be disappointed but not distraught. I've learned not to be too happy if I win nor

A grand champion certificate will be issued to rabbits that qualify. This rabbit won best of breed at a national convention show.

WHAT IT'S ALL ABOUT

There's no denying the ultimate satisfaction of winning an annual ARBA convention or the season's first local show and proving to yourself that you have outdone your fellow breeders from the nest box to the show table. It's the same thrill as that felt by a Kentucky Derby–winning stable owner and the builder of the car that captures the Indianapolis 500. For a lot of raisers, that is what it's all about. Nevertheless, all raisers, whether they show or not, should keep their stock in winning condition. And all raisers should give the shows a whirl.

too sad if I lose. The judges are all too human. If I win, the judge is one of the best. If I lose, that judge has a lot to learn. Just kidding. Seriously, the judge's verdict is not the last word. If you win consistently, over a period of time under a number of judges, you can begin to take real pride in your herd. But isolated, individual verdicts should never influence your opinion of your stock.

Never breed rabbits according to a judge's verdict. You are your own best judge when it comes to selecting the breeders in your herd. Here's an example: One September a judge accorded one of my bucks 15th place in a class of 15. Knowing the judge was not experienced with Tans, I entered the buck in the ARBA convention show 2 months later. It was judged best out of 118!

Selecting Show Stock

As I've said, when it comes to choosing which of your rabbits to enter in shows, you know best. As you breed, you will want to keep an eye on the calendar for spring and fall show dates. Try to breed for animals that will be at the top of the age limit for their classes; for example, 5½ month juniors, 7½ month 6-to-8s. Don't show seniors over 1 to 2 years, however, as they probably won't do as well as they did in their youth. Practically never will you win with a doe that has brought up a family.

Rabbit Therapy?

A GENERAL PRACTITIONER in New Jersey has some special reasons to like rabbits and rabbit shows.

"After listening to my patients complain all day," says the genial MD, "it's nice to go out back to my rabbits. They don't say a thing."

Even physicians need occasional therapy, and for this doctor it's rabbits.

"Here we are at a rabbit show on a fairgrounds 75 miles from home. There's no telephone here. Nobody can find me. My backup has to handle all the calls himself. It's really great."

But the doc also likes shows for a very special reason: It gives him time with his three children. The rabbits offer an opportunity for the family to share a lot of fun. He finds that particularly rewarding because medical demands are great, and time spent with family is often limited.

Fur and flesh condition are two prime considerations. Don't enter a rabbit if it isn't in the peak of condition, because judges will give it short shrift. Also, watch for disqualifying features, such as crooked teeth, spots of the wrong color fur, or off-color eyes or toenails. You must be familiar with the standard for your breed. The complete *Standard of Perfection* is available from the ARBA.

How Many to Enter?

You may enter as many rabbits as you like, in an effort to improve your chances of winning. I know one man who entered hundreds at a single show. Most breeders enter from two to about twenty.

You may want to enter only the very best you have, thereby holding your entries to a minimum. After all, only one rabbit is best of breed. Or you might want the judge's opinion of a great number of your rabbits or desire to amass sweepstakes points to win a breed-club trophy. It also depends upon how many good rabbits you have available and how many you can transport. This spring I entered as few as two in a distant show and as many as twenty-five in one close to home. At one ARBA convention show, I entered only four Tans, one of each color. I won three of the four colors or varieties and best of breed. My chocolate doe came in only third but produced six great litters the following year!

Watch Them in the Nest Box

Watch your young show prospects from an early age, even when they're still in the nest box. At weaning time retain the best prospects for shows, and keep them for some weeks before narrowing your choice of show candidates. Actually, you should breed only from those you feel will produce good show stock, but you will have to choose the best among the offspring.

Conditioning for the Show Table

Once you have selected your candidates, give them special attention for weeks and even months before the show. Put them in the best possible fur and flesh condition. Successful exhibitors all have their own formulas for feeding for top condition, and I'm no different. See the box below for some of the steps I take toward the trophies.

PRESHOW CARE

- Mixed horse feed with molasses is a great fur conditioner. So is wheat germ oil on top of rabbit pellets; sunflower seeds also work well.
- Milk and mangel beets plus oats will put weight on those that need it. Powdered milk is not expensive. The consideration of cost relates to how badly you want to win. If milk pays off with a big win and breeding-stock sales, it's cheap indeed.
- Brushing with a curved-wire-tine slicker brush, then with a bristle brush, will speed a rabbit through a change of coat.
- Regular feeding and a constant source of water do more, of course, to keep well-bred rabbits in top condition than anything else I know.

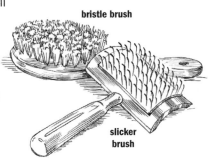

bristle brush

slicker brush

========== R A I S E R ' S   E D G E ==========

Winning Show Tips from Oren Reynolds

L ONGTIME RABBIT RAISERS know that Oren Reynolds of Illinois was the former editor of *Domestic Rabbits* magazine, a post he held for 25 years. He was also president of the ARBA at the time.

What some people may not know is that Oren was born in 1906, raised rabbits for most of the twentieth century, and was extremely successful at it. At the turn of the millennium, 98 years young, Oren was still driving his own car, going to rabbit shows — including the ARBA national convention, where he advised the board of directors — and editing the magazine six times a year. At age 100 he "retired" from editing the magazine and lived two more years, still active as an adviser to the ARBA board of directors. A butcher by trade, Oren raised New Zealand Reds in the 1930s. In 1946 he started with the Champagne D'Argent, and for a long time he was known as "Mr. Champagne" because he won more first places and bests of breed in shows than any other breeder. The last I knew he had won best of breed with his Champagnes at seventeen national conventions. And I thought I did well to win that honor just once with my Tans!

So how did he do it? Here are some of his tips:

1. Become selective. Oren said, "I am a firm believer in judicious culling. In fact, 90 percent of the mistakes made in rabbit raising are made the day the young litters are culled. The faults are there but hard to see. As the animal grows he doesn't outgrow the faults; they just become easier to see because they develop as he does. Either a youngster has a good, smooth, full rump when weaned or he will never have it."

2. Maintain only a few does. Oren never kept more than 12 breeding does at any one time. Except for an occasional doe every 3 or 4 years — usually one that carried blood from his own stock — he did not introduce new genes into his herd. Oren advised: "Once you start inbreeding you get the bad points bred out, and the only way to make sure you don't reintroduce them is to keep out strange blood. At the beginning I was pleased to get one or two keepers from a litter of seven or more. Now if I can't keep five from a litter of seven, I am displeased and consider it almost a failure. I could raise all I needed or wanted with the 12 does."

3. Breed for better fur. Oren reminisced, "It is most disappointing to be beaten show after show on fur. To produce a good balanced fur throughout the herd takes some work. I tried mating what was considered ideal fur on both the buck and doe but came away with what I considered nothing.

"After trial and error I picked does with all the texture I could get on them, even to the extent of being harsh and wiry, and forgetting the density. Even if you could see through to the hide at a glance, if the fur had texture the rabbit was used.

"On the buck, I reversed the procedure. I took bucks with all the fur density that could be obtained and forgot the texture, even to the extent of using some whose fur felt like cotton. I found that this combination would produce better fur on more youngsters."

4. Feed low-protein pellets. Oren fed whole grains and hay: "In those days I never saw an animal with fat rolls, and it seemed one could keep them in better condition longer and arrive at it sooner after a molt than is possible today. I had better luck with a 14-percent protein pellet and would feed it today except I can't find it anywhere. You could get just as fast growth on a 14-percent pellet as on any other, and it was easier to avoid fat rolls and flab. With the usual 16- or 17-percent pellet we have today, during hot weather I cut down on pellets and add whole-grain oats and barley, to reduce the percentage of protein. I reverse the procedure when it cools down."

5. Mate for best results. Oren advised, "In mating, one should have a picture of what he wants and then pick the animals that have the desired characteristics. Whether you get what you want is a gamble at best, but the odds in your favor increase when the desired characteristics have been in evidence for several generations or at least appeared now and then. Don't mate animals whose ancestors have shown the same weakness repeatedly. If you can't do anything else, mate the strength of one animal against the weakness of another."

One more thing from me. If you are a new rabbit raiser, be patient. Don't expect instant results. Oren showed his Champagnes for two years before he won a first-place ribbon.

Make the Rabbit Shows Work for You

For many years, as part of my employment, I attended a variety of trade shows in the hardware, lawn and garden, farm, publishing, investing, oil, and pharmaceutical industries. Some were small, and some were huge; for example, the National Hardware Show at McCormick Place in Chicago. This show features 3,000 exhibitors. If the 90,000 attendees were to see each of the 3,000 booths, they would have to hike 20 miles of aisles.

As an exhibitor at these shows, I learned that merely placing an ad in the show catalog or directory wasn't enough. I did that. But I also placed ads in trade journals read by people who would attend. In these ads I made a specific offer of a "show special" to be available at the show. It could be "one free with a dozen," or some similar offer.

The ads really only reinforced my best method of selling at the show, which was direct mail. For a fee I would obtain a list of people who had either indicated they were planning to attend the show or had attended the previous year's event. Then, before the show, I would mail each of them a letter outlining my special offer, obtainable at the show or even before.

How does this experience apply to rabbit shows? If you've studied the marketing ideas in chapter 7, you'll be able to guess that placing an ad in a show catalog listing specific rabbits at specific prices instead of the business-card type of notice will help you sell at a show. If you obtain the names and addresses of all those who have been sent a catalog, you can mail out a show-special offer. You may be able to obtain this list from the show secretary, or the club sponsoring the show might sell you the list. It would be a good way, in fact, for clubs to help cover some of their show costs. You might want to suggest that at your next club meeting: In the show catalog, for instance, the club might place a notice offering the list of catalog recipients to any and all at a reasonable price, such as $5 to $10 per one hundred names or more if the names come already on labels. If you can't get such a list from the club, put together your own list of breeders in the vicinity who are likely to attend a given show.

When you mail out the special offer, ask that potential buyers reserve any animals they want with a deposit in advance of the show. You can

R A I S E R ' S   E D G E

A Show Solution from Maine

D EBBIE VIGUE, a Maine accountant, has been active in rabbits for years and came up with a solution to a common malady at rabbit shows: writer's block.

It has always puzzled me that rabbit-show judges get paid for their work from entry fees, but the writers, or recording secretaries, get nothing. No wonder it's hard to find volunteer writers during a day of competition and fun. Like many others, I never tried to rectify the situation. But ingenious Debbie found a way to find willing writers without paying them.

"People step forward to volunteer for the job now," she said. "The solution was easy — offer an incentive. The first time we did, we were rewarded with immediate success.

"Raffle tickets were given to writers for a special 'writers only' raffle. Tickets could not be bought. They had to be earned by writing for the judges. Our first item offered no doubt had an impact on the success. It was a half-dozen Maine lobsters packed in a cooler all ready to take home for a ready-to-cook supper after the show. People told the show superintendent they wanted to write whenever someone put a pen down at a judging table. People actually were disappointed when they did not get a chance to write. Several people wrote all day, never budging. We call it 'lust for lobsters.'"

Writer raffle items offered since then have included carrying and hanging cages, lamps, wall hangings, director's chairs (great for resting at a show), food baskets, sweatshirts, and even Maine crabs.

"I hope this idea will be put to good use throughout the country," she said. "When you see something that needs to be improved at a show, stop complaining about it and work to find a solution."

promise to refund it if they change their minds. Then they go to the show with money earmarked for your animals. You might go home without any rabbits but with a pocketful of dough.

This procedure takes some work on your part and, like any small business, requires a modest investment for stationery and postage, as well as an investment of your time. But think of it: The show brings together a lot of potential buyers. It's great if they already have you and your animals in mind when they arrive. You will stand out from the crowd, and the time and money you've spent will be repaid.

Rabbit shows are a great source of fun and pride for all involved.

11

Rabbit Associations and the Future of the Industry

HUNDREDS OF RABBIT ASSOCIATIONS exist in the United States. They benefit their members in several important ways at the local, state, regional, and national levels. In addition, they offer their members the opportunity to advance the cause of rabbits in general and favorite breeds in particular.

Although these organizations are primarily fancier-focused today, a time may come when they or other associations offer greater support and advocacy for raisers who sell their animals for human consumption. Because rabbit is nutritious and delicious, the demand for rabbit production, raisers, and associations that promote the sale of rabbit meat may be on the horizon.

National Organizations

The American Rabbit Breeders Association, which maintains a full-time, paid secretary and an office staff in Bloomington, Illinois, is the parent organization. Affiliated with the ARBA on the national level are breed-specialty clubs dedicated to the advancement of the individual breeds. There are also regional, state, and local associations affiliated with the ARBA.

The ARBA holds an annual meeting and convention show each year, usually in October or November. Members enter many thousands of rabbits in this weeklong show, the biggest of all United States rabbit competitions. More than 20,000 rabbits may be exhibited.

The breed-specialty clubs also conduct meetings at these annual conventions. They usually hold a national show devoted exclusively to their breed at some other time of the year. The rest of the year, the ARBA and the breed-specialty clubs maintain communication with their members through *Domestic Rabbits* magazine and the specialty-club newsletters. I'm especially familiar with this procedure, having founded *Domestic Rabbits* and served as its first editor and having edited the newsletter of the American Tan Rabbit Club.

Regional Groups

At the regional, state, and local levels, club meetings are more frequent, usually once a month for local clubs. Members assemble at a central location to discuss various aspects of rabbit raising. Rabbits are bought and sold, and tidbits of information on marketing, breeding, management, and every other imaginable topic are exchanged. Each local, state, or regional association conducts one or two shows per year, which requires much planning and work by volunteers from the ranks. In addition, newsletters, e-mails, and Web sites keep members informed about club activities, including annual social events, such as picnics or dinners, as well as tips and updates on rabbit raising.

Every successful rabbit raiser I know maintains close contact with other raisers in the various associations because of the invaluable exchange of information that takes place. What is difficult to make clear to some breeders is that membership fees license them to help the association; the fees don't entitle them to any largesse. These are amateur, volunteer associations of individual breeders working together for the common good at each level. With the exception of the ARBA staff, part-time paid secretaries in a few organizations, and the paid workers in the processing plants of a handful of commercial rabbit cooperatives scattered across the country, rabbit-association members are amateur, volunteer workers.

ARBA Benefits

Members of the ARBA are entitled to register rabbits, earn grand championship certificates, vote for officers and directors, and generally have a voice in the direction the rabbit industry is taking. Membership includes a copy of the 200-page *Official Guidebook to Raising Better Rabbits* and a subscription to *Domestic Rabbits* magazine. Members also receive the *Yearbook*, which is a directory of all the members in the United States and abroad and also contains the constitution and bylaws of the association. A raiser who is not a member is not up-to-date on what's happening in the world of rabbits. I can't afford not to belong, and neither can the other 30,000 members.

To join the ARBA, visit www.arba.net or write for an application (see page 207 for mailing address).

Into the Future

Despite the several hundred rabbit raisers' associations that exist in the United States, there is still a need for another. Except for a few local commercial cooperatives, there is no national group dedicated to the promotion of the rabbit for what it is best suited: human consumption.

There is a commercial department within the ARBA, but mere volunteers in the commercial department haven't the ability or the equipment to promote rabbit meat, and unfortunately, neither do the leaders of the ARBA. The ARBA and affiliated organizations devote 99 percent of their effort to promotion of the fancy side, or the showing, of rabbits because that's what its board of directors wants it to do. That is not all bad, because it is from the fanciers that commercial rabbit breeders obtain their foundation stock.

Increasing the Demand for Meat

Despite the importance and value of the ARBA focus on showing rabbits, it strikes me that the best single goal of the ARBA would be to increase the American consumer's demand for rabbit meat. The way to do that

A PROFITABLE RABBIT MARKET

What would happen if rabbit meat found its way to the American dinner table on a regular basis?

- More rabbits would be raised.
- More breeding stock would be needed.
- More rabbit feed would be produced, and volume production would affect the price of feed favorably.
- More wire hutches, feeders, watering equipment, nest boxes, and sundry other types of equipment would be manufactured and sold, and the price for all of these items would be affected favorably by volume manufacture and sales.
- More animal-health companies would research, clear with the Food and Drug Administration, and market health and nutrition products for rabbits. That would make the task of keeping rabbits healthy and productive easier; medicine for rabbits would become widely available where it currently is practically nonexistent, and veterinarians would have more interest, information, and expertise in rabbit medicine.
- Government, university, and industry research into rabbit breeding and management would increase. Thus rabbit raising would become a more exact and more recognized agricultural science. All engaged would derive more pleasure and more profit.

The ultimate beneficiary would be the consumer, who would find delicious, nutritious rabbit abundantly available to provide a welcome mealtime change of pace at a competitive price.

would be to start still another organization, one that might be called the American Commercial Rabbit Council. Inasmuch as there is practically no demand for rabbit meat in the United States compared to that of other meats (per capita annual consumption of beef in the United States is about 67 pounds [30.4 kg]; of rabbit, 4 ounces [113.4 g]), this demand has to be created. Critics of the idea say there is not a large enough supply. Of course, there is only one answer to such a chicken-and-egg question. There is only one thing that creates a *supply* of anything. That is, quite simply, *demand*. An example is any fuel-efficient car: when gasoline prices go up, or people think they will, so does the demand for such vehicles.

It seems to me that any person who believes rabbits are mainly for eating (as I do) would agree that the half-dozen points I give in the box on page 188 are desirable and reasonable. Even those who are strictly fanciers (and I am a fancier, as well, after all) would have to agree, I think, that at least some are desirable. For example, with greater need for breeding stock, fanciers would find an expanded market awaiting their production and could enjoy lower feed prices, more and better equipment, and advances in health and management practices. Part-time breeders who always wanted to expand into full-time operation would have a sufficient market to warrant the move. Others would be able to breed more often, being better assured of demand.

A Promotional Campaign

Demand is the key word. Could such a demand be created? I am sure of it. A sustained program of publicity and promotion in consumer media could do the job. Florida oranges, California prunes (now labeled "dried plums"), and catfish are examples of agricultural products that have benefited from such promotion. Beef, our favorite meat, receives several million dollars worth of promotion every year. But rabbit, known to but a few, gets next to nothing. The Pel-Freez Company (Rogers, Arkansas), which sells frozen rabbit meat nationally, takes small ads in some women's magazines from time to time. Except for a few hundred dollars in bumper stickers and similar materials for National Rabbit Week, which occurs in the summer — a very poor time to promote — the ARBA spends little.

An organized and professionally executed publicity and promotion plan would achieve the goals I mentioned and more. It would be good for rabbit raisers and consumers. But it would be terrific for milling companies, equipment suppliers, laboratories, processors, and commercial breeders. Quite simply, they would make more money.

How Would a Campaign Work?

Here's where the ARBA could come in. The organization could unite those who supply the raisers, in a special unit that might be called the

American Commercial Rabbit Council (or something of that sort). This council would hire a public-relations agency to promote rabbit meat at a rate that would keep demand just ahead of supply.

Such a campaign would begin in regional media, at a moderate pace, and eventually expand into national media, increasing demand at a rate that would give breeders and processors time to tool up to meet it. In the case of rabbits, of course, that wouldn't take too much time because of the animal's renowned reputation for fast maturity and reproduction.

It would be the task of the American Commercial Rabbit Council to gather members, who would be required to hold ARBA membership. The council would also solicit funds and spend the money wisely to increase the demand for rabbit meat.

Finding Labor and Funds

Who would do the work to form such a council? I have discussed this idea with three trade association agencies, all of which have expressed a great deal of interest. One has submitted to me a detailed plan. Another is already successfully operating such a council in another branch of the food-supply sector. All three are successful firms ready and willing to go to work.

This work can't be done successfully by part-time volunteers, such as those who run the various rabbit-fancier organizations. It requires full-time professionals with a profit motive.

All this would take a little seed money, but there are several ways to raise those dollars. The most lucrative solution is to solicit large industries and appeal to their profit motive. For example, small raisers, who may produce no more than three hundred rabbits per year, pay dues of $20 each per year to belong to ARBA and get more than their money's worth. But the ARBA dues for the nation's largest suppliers of rabbit feed are still $20. What if the ARBA was actually helping these suppliers sell this feed by expanding the demand for rabbit meat? Certainly, dues for such large suppliers could exceed $20, up to perhaps several hundred or a thousand dollars or more per year. One corporation that belongs to ARBA at $20 per year also belongs to several other livestock organizations. To one of these, it pays dues of $50,000 per year! Of course, it gets its money's worth, or it wouldn't pay the price.

A Show of Support

I proposed my idea to the members of the ARBA through articles in *Domestic Rabbits* magazine. More than one hundred members responded with letters, and many others have spoken to me about it at rabbit shows. All but four liked the idea, and three were processors who feared competition. Here's what a few of the ARBA members have told me:

I'm behind you 100 percent. Thank you for some great articles. — Florida

Ribbons don't pay the grain. I say yes, spend the money, and help us sell meat. — Massachusetts

Man, am I glad someone has finally found the front end of the problem in the rabbit industry. Your idea is just great. — Florida

My feed company here said it would get behind something like this, too. Many would profit, not just rabbit raisers. — Minnesota

I do not have a lot of New Zealands but would expand if I could be sure of a market for them. — Iowa

Go get 'em, boy. We've been messin' around too long. — Oregon

The soundest proposal I've heard of. Every rabbit breeder stands to gain. We need this. There is no way to lose. — Montana

I don't believe ARBA could use its funds for a better purpose. — Kansas

If we don't promote rabbits in a good professional way, we will never get to first base with them. — Minnesota

I sure hope this will be the year to promote the rabbit. — Louisiana

I am all for getting a market for rabbits. I have fifty I want to get rid of now. No sale! — Michigan

I joined the ARBA in the hope that it would promote the commercial aspect of this business, and I think we're finally seeing the light. — Michigan

To me this is one of the very best ideas that has yet come forth for the rabbit industry. — Kentucky

Other livestock organizations really promote. Why not the ARBA? Take the money and go. — New York

Promote the food as much as the fancy, the rabbit more than the association. — Georgia

This idea will help the commercial grower better than any other idea. — Missouri

Full steam ahead. We're 40 years behind now. — Illinois

I live in an area that has a lot of publicity for pork and beef, and I can readily see the sound sense in this project. — Nebraska

Public relations is the answer. I stand behind you. — Tennessee

I sell to Pel-Freez. I think your idea is a good one, and I support you 100 percent. — Oklahoma

This would be good for all the breeders in the country. — Oklahoma

Let's start a massive campaign to get Americans to start eating rabbit meat. — Connecticut

The Rabbit's Potential

And so the responses went. But while the membership of ARBA seems overwhelmingly in favor of this idea, the leadership is not, preferring to remain strictly a fancier organization. Still, the rabbit holds great possibilities as a meat animal for Americans, but no nationally successful industry can exist without national promotion. It seems obvious that in time the ARBA either has to organize a bona fide commercial unit or find that someone else has come along and done it.

Perhaps the main reason that a single goal of rabbit-meat promotion has not yet been sought is simply because the rabbit has so many diverse uses. It is so versatile — just like the Shmoo to which I alluded at the beginning of this book. And thus it is difficult to get agreement to work toward a single goal. Those of us who have rabbits see them in so many different ways that we can't yet see eye to eye on promoting them. But one of these days, I believe, we will.

WHAT'S IN A NAME?

For years I agreed with those who think we should come up with another name for rabbit meat, an alternative to what we call it now, rabbit. After all, it's pork, not pig; beef, not cow or steer; even, occasionally, poultry instead of chicken. These names are derived from the French, so you might say we ought to call rabbit meat lapin. Meanwhile, to promote their favorite meats, in national ad campaigns, svelte skater Peggy Fleming called pork "the other white meat" and actor James Garner said beef is "real food for real people."

It's true that some people can't get the name past their palate; others want to be fooled into thinking a euphemism will protect their arteries from an overabundance of cholesterol. The United States is loaded with people who think meat originates in plastic wrap in the back of the supermarket. You tell them they are eating rabbit, and all they can think of is cute Easter bunnies, Peter Cottontail, Beatrix Potter's Peter Rabbit, and Bugs Bunny, or else they believe they're consuming a wild, tough, dry, stringy rodent in a hobo stew.

I used to think that it needed a new name, but I've thought about it some more, and now I don't believe rabbit needs any other name. Here's why — steer, pig, and chicken are ordinary. They need a special name to make them palatable. Note that pheasant is so classy, it doesn't need another name. Nor squab. Not even duck or turkey. If you are already special, you don't need a special name, even if you are an odd-looking culinary creature or a less-than-comely comestible such as squid, clam, crab, lobster, or oyster.

I could go on with this, but you're getting hungry and need to get to the next chapter where you can start cooking and eating rabbit. Sure, we need more education about the delights of rabbit meat, and that means promotion to get more people to try it and be convinced. But let's stick with the name that everyone already knows.

12

Cooking Rabbit

ONE TASTE WILL CONVINCE YOU that more domestic rabbit should be eaten. A whole lot of people are missing out on a good thing. It's not easy, however, to get some people to take that first bite.

My young children would *never* eat rabbit. Perish the thought. They did eat a lot of chicken, and occasionally asked where, perchance, the wings might be and how come there were extra drumsticks. My reply, of course, was that should a person care to purchase more of certain chicken parts, one can do so, snubbing wings, backs, and necks. Thus reassured, they took another bite. Of rabbit.

Domestic rabbit meat is fine-grained and pearly white. It does not taste wild or gamey, as some believe. The domestic rabbit is no more wild game than a Black Angus steer or a Rhode Island Red rooster.

Before you start raising rabbits for meat, you certainly should taste it. So I suggest you purchase a Pel-Freez fryer-broiler, available at chain supermarkets across America in the frozen-food case. The price might surprise you. It's not cheap. Usually, it's about the same price as sirloin steak. The price has a lot to do with the feed/meat conversion ratio, which we discussed earlier, plus the costs of processing, packaging, and shipping, not forgetting a profit for the grower and the processor and the retailer.

Chicken Is Cheaper

A lot of people compare rabbit to chicken, and of course, the price per pound of chicken is much lower — two-thirds to three-fourths lower. That is due partly to the absolutely amazing performance of the American poultry industry. Chicken actually costs less these days than it did 50 years ago. Per capita consumption has climbed as the price has declined. (Beef consumption simultaneously declined, mostly because Americans have become cholesterol conscious). How many items on your grocery list cost less than they did 50 years earlier? I know of no others, although in spite of the gripes of the uninformed, food remains quite a bargain in this country, taking a smaller percentage of our income than it does in any other country in the world and even a smaller proportion of our paycheck than it did years ago.

Comparing Rabbits and Chicken

The *New York Times* published a series of quotes comparing the taste of several food items to chicken. Rabbit certainly made the list, because many people say rabbit tastes like chicken. Here's what else people think tastes just like chicken: snake, frogs' legs, pheasant, alligator, iguana, tempeh (a soy cake), turtle, mako shark, and nutria. Given the choice (and I've tried all but iguana and nutria), I'll take rabbit any day.

Rabbit Lags Behind

The tortoise-paced rabbit industry will have to make some giant hops to catch chicken's feed/meat conversion ratio of about 2:1. Compare that to 4:1 for rabbit. Chicken's conversion ratio has been largely brought about by geneticists who have not yet turned their talents over to the rabbit.

But rabbit has finer bones than chicken, and the meat has a finer grain and a chewier texture, so a little fills you up. The United States Navy recognized that many years ago when it served rabbit and allotted each sailor a 6-ounce (170 g) portion, compared to 12 ounces (340 g) required for a portion of chicken. And it should be noted that rabbit-feed pellets are only about 50 percent grain; the other 50 percent is forage products, or hay, largely unused for human consumption. That means that compared to other livestock, rabbits do not compete as much with humans for food products.

High-Protein, Low-Calorie Rabbit

In addition, domestic rabbit is extremely low in calories and high in protein content.

Here is a U.S. Department of Agriculture (USDA) statistical breakdown of several meats:

Meat	Percentage of Protein	Percentage of Fat	Percentage of Moisture	Calories per Pound
Rabbit	20.8	10.25	27.9	795
Chicken	20.0	11.00	67.6	810
Veal (medium-fat)	18.8	14.00	66.0	910
Turkey (medium-fat)	20.1	22.20	58.3	1190
Beef	16.3	28.00	55.0	1440
Lamb (medium-fat)	15.7	27.70	55.8	1420
Pork (medium-fat)	11.9	45.00	42.0	2050

You can see from this table that rabbit has more protein and fewer calories than any of the popular meats! It also contains less moisture, which means you don't purchase water. Neither do you pay for the skin, which, in the case of chicken, many find fatty and rubbery and only throw away. Recent studies have also shown that rabbit compares favorably to chicken in its low cholesterol content. Anyone concerned about heart disease should consider the merits of rabbit meat.

Preparing for Cooking

After I butcher a fryer-broiler, I chill the meat for a couple of hours or overnight in the refrigerator, cut up the pieces, and rinse them thoroughly in cold water (see Butchering a Rabbit in chapter 7). I don't soak

IT'S SO DIGESTIBLE

Rabbit is easy to digest. Remember that when feeding children, senior citizens, or those with weak stomachs or digestive problems. Those on bland, soft-food diets take well to the tender texture and mild flavor of domestic rabbit.

them in water as some people do. Then the rabbit goes into the freezer. Ordinarily, all my meat rabbits get a stay in the freezer. My wife, Alice, says a rabbit stays there until we forget which one it is. We never eat a rabbit we know personally.

Freezing

You can freeze uncooked rabbit, washed and chilled, whole or in parts. Wrap in moisture- and vapor-resistant material suitable for freezing. Alice uses plastic freezer bags, but you can use heavy-duty aluminum foil or freezer paper. Be sure to squeeze the air out of the package before sealing. It's also a good idea to mark the date on the package.

Prepare cooked rabbit for freezing the same way, unless you want to include gravy or sauce. In that case, pack it in rigid containers with tight lids. In our house, though, we never freeze cooked rabbit. Somebody eats it up first. Fresh rabbit will store in the freezer for 6 months; cooked rabbit, about 2 months.

Thawing Rabbit

It's best to thaw rabbit in the refrigerator, with the wrapping loosened. Pieces will thaw in 4 to 9 hours; whole rabbits may take 12 to 16 hours, and the bigger the rabbit, the longer the thaw time. To thaw faster, plop the rabbit, bag and all, into cold water. Don't refreeze either cooked or uncooked rabbit once it has been thawed. Successive thawing and refreezing lowers quality and can pose food poisoning risks. When it comes to storage, keep in mind that an entire rabbit or two make a meal for the whole family. If you butcher a hog or a steer, you have a storage problem. But rabbits store very nicely "on the hoof" in the hutch. You can pick one out and cook it up, and it's gone at one sitting.

The Recipes

Despite all my disparaging comments about chicken, remember that you can cook rabbit using almost any chicken recipe. And you can also prepare some veal-type dishes with rabbit. You can cook it fancy or plain — any way you like. The following are just a few samples of the hundreds of recipes you can try.

Alice Bennett's Crispy Oven-Fried Rabbit
(Our Children's Favorite)

With this recipe, you'll make "fried" rabbit without all the grease. It's a prizewinner, having garnered top honors at a statewide 4-H cooking contest in Texas one year.

> 1 fryer-broiler rabbit (2–2½ pounds
> dressed), cut up
> 1 egg, well beaten
> 2 cups potato chips, finely crushed
> ¼ cup butter or margarine

1. Dip rabbit pieces in beaten egg, then coat with potato chips.

2. Melt butter in a shallow baking pan.

3. Arrange rabbit pieces in the prepared pan and bake in a preheated 375°F (190°C) oven for 30 minutes.

4. Turn the pieces and bake for another 30 minutes, or until well done. Test with a fork. The meat should be tender and juicy.

SERVES 4

DID YOU EVER TASTE "MARSH RABBIT"?

Often I hear from people who can't get past the idea of eating domestic rabbit meat, yet there are plenty of diners who think "marsh rabbit" is delectable.

What is marsh rabbit?

In Delaware that's what they call muskrat, which happy eaters order up with side dishes of stewed tomatoes and fried potatoes. Trappers get about $8 for the pelts, and restaurants can't get enough skinned muskrats. In one 3-month period, six hundred trappers in Delaware harvested 140,000 muskrats, enough for 14,000 full-length coats, 140,000 dinners, and vast amounts of musk oil, which is used in perfume.

Golden-Brown Fried Rabbit

If you like fried chicken, you will love fried rabbit.

½ cup all-purpose flour
1½ teaspoons salt
½ teaspoon pepper
1 rabbit fryer, cut up
2 tablespoons olive oil

1. Mix the flour, the salt, and the pepper in a plastic bag.

2. Add the rabbit pieces and toss to coat well with the flour mixture.

3. Heat the olive oil in a heavy skillet.

4. Place the rabbit pieces in the skillet over high heat, turning to brown evenly on all sides.

5. Reduce heat to low, cover, and cook slowly, 40 to 50 minutes, or until tender. For a crisp crust, uncover for the last 10 to 15 minutes.

SERVES 4

For a great marsh-rabbit dinner, cut off the heads and tails, skin the muskrats, and soak them in salted water to draw out the disagreeable wild taste. Parboil the meat with onions and seasoning for about an hour, brown it in an iron skillet with sage, salt, and pepper. When done, the meat is dark and stringy and can be served up in a casserole dish. No, this is not one of those jokes where you throw the muskrat out and eat the dish.

On the other hand, maybe you just want to forget about it. Why dine on marsh rabbit when the real thing is so delicious? The real McCoy is what appeals to me.

Braised Rabbit

Alice likes to serve this dish in the summer with fresh vegetables from our garden. Son Bob contributes the thyme from his herb garden near the back door. He says that's why it tastes so good.

½ cup flour
1½ teaspoons thyme leaves, crumbled
1 teaspoon salt
½ teaspoon onion salt
½ teaspoon celery salt
⅛ teaspoon pepper
2 rabbit fryers (about 2 pounds each), cut up
¼ cup olive oil
1 cup water

1. Put flour and thyme, salt, onion salt, celery salt, and pepper in a plastic bag.

2. Add rabbit pieces and toss to coat well with flour mixture.

3. Heat olive oil in skillet.

4. Brown the rabbit in the olive oil, turning to brown on all sides.

5. Add water, cover, and simmer until rabbit is tender, for about and hour, or a little longer if the pieces are large. Test with a fork for tenderness.

SERVES 6 to 8

Note: You can thicken the cooking liquid with flour for a delightful gravy.

Rabbit Casserole

This recipe takes more time to prepare than many, but some people think it's well worth the wait.

> 4 slices bacon
> 2 rabbit fryers (about 2 pounds each), cut up
> 3 medium-size onions, quartered
> 2 green peppers, chopped
> 1 clove garlic, peeled and crushed
> ½ cup dry white wine
> 1 can (16 oz) whole tomatoes
> 1 can (8 oz) condensed cream of mushroom soup
> Salt, to taste
> 1 teaspoon marjoram, crushed
> 1 teaspoon thyme, crushed
> Rice or noodles as accompaniment

1. Sauté the bacon in a skillet until crisp. Drain on paper towels and set aside.

2. Brown rabbit a few pieces at a time in bacon drippings, then arrange in a 10-cup baking or casserole dish.

3. Add onions, green peppers, and garlic to the skillet. Add the wine, followed by the tomatoes.

4. Cook the mixture in the skillet, stirring and crushing the tomatoes until the mixture is slightly thickened, about 5 minutes.

5. Stir in the cream of mushroom soup, salt, marjoram, and thyme.

6. Heat to boiling, stirring frequently. Spoon over the rabbit in the baking dish and cover.

7. Bake in a preheated 350°F (177°C) for 1 hour or until tender.

8. Just before serving, crumble the reserved bacon and sprinkle it over the rabbit and vegetables. Serve with hot buttered rice or noodles.

SERVES 6 to 8

Rabbit Cacciatore

If you like Italian restaurants, you might stay home and prepare this dish, which is difficult to find when dining out.

    1 rabbit fryer, cut up
    1 large onion, chopped
    1 can (6 oz) tomato paste
    ½ teaspoon garlic salt
    ½ teaspoon oregano
    2 bay leaves
    ½ teaspoon sugar
        Salt and pepper to taste
    8–9 ounces spaghetti

1. Put rabbit in salted boiling water to cover and simmer until tender. Remove meat from broth. Set the broth aside.

2. Remove meat from bones, if desired.

3. Add the onion, tomato paste, garlic salt, oregano, bay leaves, sugar, salt, and pepper to the rabbit broth and simmer for 1 hour.

4. Cook spaghetti.

5. Return rabbit to the reduced broth. Serve on a bed of spaghetti.

SERVES 4

OTHER RECIPE IDEAS

You can find a recipe for hasenpfeffer, a highly seasoned rabbit stew, in almost any cookbook. And you can barbecue rabbit in the backyard just like chicken. I like to parboil it first, cook it inside some aluminum foil with the sauce, and then brown it while basting with more sauce, right over the coals. You can also smoke it if you like. (You can buy smoked rabbit in some places or through the mail for about $20 a pound!) I'm sure all good cooks will get the idea that rabbit can be prepared in any number of delicious ways, fancy or just down-home.

AFTERWORD

WHAT KINDS OF PEOPLE RAISE RABBITS? Are any of them like you? Answers to these questions may help you envision yourself as a successful rabbit raiser. Take my New Jersey friend, a family man in his 30s, with a position in the purchasing department of a local firm. Only a couple of years ago, he started to raise a few rabbits for his own table in a neat little storage shed in his suburban backyard. His goal is to put a rabbit on the table each week. Not only has he reached his objective, thereby making a dent in his meat bill; he also sells breeding stock to breeders and extra fryers to a butcher. And he is a frequent winner of blue ribbons at regional rabbit shows. His success arrived quickly because he saw the necessity of starting right. He purchased foundation stock from the nation's top producer of his chosen breed. Then he housed, fed, and managed these rabbits according to the most modern methods.

I have another friend in Vermont, about the same age, a professional in a utility company, the father of two school-age youngsters, and a resident of one of the larger towns. He has raised rabbits for several years. He has always insisted upon the very best rabbits available and kept his rabbitry, which is in one-half of a detached two-car garage, scrupulously clean. People actually wait for his rabbits to be born to purchase them for their own rabbitries. But his standards are so high that only the very best of his production is sold for breeding stock. The rest go for meat. He slaughters some for his own table because his family really enjoys rabbit.

But they also like a good steak once in a while. So, in a satisfying arrangement for all concerned, the Vermonter swaps dressed rabbit meat, pound for pound, for beef, with the operator of a freezer locker. His friends and neighbors know he raises rabbits, but I call him a cattleman.

A professor of animal husbandry at the University of Maine pioneered the importation of my favorite fancy breed into the United States from England and Holland. With careful mating and study over many years, he steadily improved this breed. One year the association that sponsors the breed in this country voted him their most outstanding member and presented him with a handsome plaque to prove it.

A Long Island airline pilot and his family are also avid fanciers. And both a New Jersey dentist and a waitress lead 4-H clubs dedicated to raising rabbits. The waitress has written two fine manuals and the dentist shows his charges how to build the most advanced rabbit housing.

Here are other rabbit raisers: A pharmacist produces laboratory stock in Puerto Rico; so do a factory worker in Maryland, a carpenter in Pennsylvania, and a farmer in New Hampshire. A New York City marketing executive raises breeding stock, show stock, and meat rabbits at his suburban home. And he still has plenty of time to play golf. An Oregon family — he's a government administrator; she's a weaver and the head of her own mail-order business; their daughter is a high-school student — raise many fancy breeds successfully, fill orders for breeding stock from all over the country and Canada, and take home trophies. At one show they won the top trophy for each of ten different breeds!

A monk keeps his rabbits behind the monastery garden. Unfortunately, his rabbits are so handsome they sell at a nice profit and have compromised his vow of poverty. A minister is losing money on rabbits but doesn't mind a bit. He heads a home for orphaned boys. He bought them rabbits and hutches to teach them animal husbandry and business principles. So far, they are still in the red. But the minister is all smiles. Those boys are really busy, and he loves it.

A surgeon, a chiropractor, a lawyer, an ex-convict, and a man who has sewn the bristles into brushes for 30 years are all successful rabbit raisers. Two Boy Scouts recently spent several months raising rabbits to earn a merit badge. I served as their counselor, after devising the requirements and writing a book on rabbit raising for the Boy Scouts of America.

A few years ago a young woman in New York State learned how to hand-spin wool. Now she raises Angora rabbits and spins their wool into yarn. Her abilities are so highly respected that she teaches hand spinning, dyeing, and weaving at two colleges, two art centers, and a museum.

You may be wondering if it is possible to really make it big with rabbits. Sure, it's possible, and the chances are improving each year, but it takes a special person to make it big at most anything. And the production of livestock is one of the most challenging fields anyone can enter.

One of the most exceptional people ever to raise rabbits was Edward H. Stahl, who died in 1973 at the age of 87. I had the privilege of knowing Ed Stahl, who preceded me as an ARBA publicity worker and gave me a lot of guidance. I consider Ed Stahl to be the single most important person in the development of the United States rabbit industry.

Ed was born in New York State in 1886. At the age of 20, he left home and hired on as a deckhand with a tramp steamer. It was scheduled to go to New Orleans, but he left the ship early and rode a freight train to Kansas City, Missouri. In nearby Holmes Park he got a job in a foundry, married, and bought two acres behind the foundry. A coworker offered Ed a pair of rabbits. He housed them in a rundown corncrib in 1913, and by 1928 he had spent so much time with them that his boss told him he had to make a choice between the job and the rabbits.

"I chose rabbits," Ed said, having built a rabbit business that was grossing $350,000 per year. At one time he had 65 full-time employees and a second rabbitry in New York State. He sold breeding stock through the Sears, Roebuck catalog and in a single month sold $49,000 worth. It is said that Ed Stahl, with his rabbit sales and associated activities, including publishing and the manufacture of supplies, became a millionaire (back when a million dollars wasn't chicken feed). Is that what you mean by really making it big with rabbits?

Many Americans have achieved success with rabbits because the rabbit has so much going for it. But it is perhaps because it looks so easy to raise rabbits that many of those who try it become discouraged and fail in a very short time. One vital aspect lies hidden beneath that soft and furry exterior — the necessity for starting right and raising them right.

READING ABOUT RABBITS

A GOOD LIBRARY IS ESSENTIAL for the person or family raising animals. No one can remember all of the information this endeavor requires, and a good library will provide it at your fingertips. New ideas, techniques, and theories are always being put forth, so keep your library card up to date. There are many good books available; here are some that are excellent choices.

The following list of books, pamphlets, and magazines includes only those that I have read myself (or written) and found useful.

Books and Pamphlets

American Rabbit Breeders Association. *Official Guide Book, Raising Better Rabbits and Cavies.* Bloomington, IL: ARBA, 2000. Two hundred and forty pages of essential information for all rabbit raisers. Written by amateur volunteers — consequently, much is difficult reading — but nevertheless a fine compilation of lore from years of experience. Free with membership.

Bennett, Bob. *Rabbits . . . as a Hobby.* Neptune City, NJ: TFH Publications, 1992. Completely illustrated with full-color photographs taken by the world's best photographers of rabbits and beautifully printed. Text limited to choosing and caring for a pet.

———. *Build Rabbit Housing: A Storey Country Wisdom Bulletin (A-82).* North Adams, MA: Storey Publishing, 1983. Thirty-two illustrated pages of instructions for building rabbit hutches, nest boxes, and other equipment. Updated information, 1988.

Shields, Earl B. *Raising Earthworms for Profit.* Eagle River, WI: Shields Publications, 1994. Best little book on earthworms I've read. Special chapter on worms and rabbits.

Magazine

Domestic Rabbits
American Rabbit Breeders Association
309-664-7500
www.arba.net
Bimonthly. Excellent articles. Free with ARBA membership.

Newsletters and Guidebooks

For each recognized breed, there is a newsletter and a guidebook produced by the specialty club devoted to the particular breed. Club membership is required to receive these publications, but there are additional benefits to be gained. Addresses for these breed-specialty clubs and for hundreds of state, regional, and local breeder associations, many of which also publish newsletters and maintain Web sites, can be found in the ARBA *Yearbook*, which is included in ARBA membership.

Rabbit and Equipment Suppliers

Rabbit equipment is now available by mail-order, online, and at many feed stores. If you can't find what you need at the store, ask the manager to order it from a distributor.

For a list of current suppliers of many breeds of rabbits, as well as equipment and an ARBA application form, send a self-addressed stamped envelope to Bob Bennett, 133 Governors Lane, Shelburne, VT 05482.

You may also contact Bob by e-mail at rabbitalogue@hotmail.com.

BREED GALLERY

RABBIT BREEDS FALL INTO FOUR CATEGORIES by size: giant, medium, small, and dwarf. Fur types are also four: normal, rex, satin, and angora. What follows is a brief description of several breeds, along with a few comments that should help you as you consider what breed or breeds to raise.

Each year the show at the American Rabbit Breeders Association National Convention draws thousands of entries. The most recent show included more than 23,000 rabbits!

To give you a sense of the types of rabbits that are popular among raisers in the United States, here's a list of the most numerous and popular breeds at the show. The rankings and the numbers of rabbits entered really don't vary much from year to year, although a few breeds enjoy short-lived popularity.

REMINDER

The best breed to raise is one that somebody else will buy, unless you truly wish to be a rabbit *keeper*. In addition, it's best to stick to just one or two breeds in order to make the best use of your time, to become really good at breeding, to streamline your record-keeping chores, and to use hutches and equipment efficiently.

The New Zealand White doubtless is the most popular breed because that's the one that most people buy, although many of the smaller breeds are better represented at most rabbit shows. Recall that most meat rabbits are New Zealand Whites; also many are Californians.

GIANTS

Checkered Giant

IDEAL WEIGHT: 12 pounds (5.4 kg)

VARIETIES: Black, Blue

FUR TYPE: Normal

MARKET: Less desirable than the other giants

PROBLEMS: Boney, rangy bodies; active movement burns pounds of feed but does not produce much meat per pound; requires a big hutch

GIANTS

Flemish Giant

IDEAL WEIGHT: 15 pounds (6.8 kg)

VARIETIES: Black, Blue, Fawn, Light Gray, Sandy, Steel Gray, White

FUR TYPE: Normal

MARKET: Practically none, although this is ostensibly a meat rabbit

PROBLEMS: Big eater and too boney; needs a very large hutch — in fact, requires a hutch much bigger than any available commercially

Giant Chinchilla

IDEAL WEIGHT: 13–14 pounds (5.9–6.4 kg)

VARIETIES: Chinchilla only. The breed standard requires that the coat resemble a real chinchilla, with the undercolor being a dark slate blue, the intermediate portion of the hair shaft being light pearl, and the top edge being a narrow black band; a wavy ticking tops it all.

FUR TYPE: Normal

MARKET: Not a much better situation than that for Flemish Giant

PROBLEMS: Like Flemish Giant, a big, boney eater that needs a huge hutch. A better choice if you like chinchilla fur would be the American Chinchilla (10–11 pounds [4.5–5.0 kg]) or the Standard Chinchilla (6 pounds [2.7 kg]).

MEDIUMS

Californian

IDEAL WEIGHT: 9 pounds (4 kg)

VARIETIES: One

FUR TYPE: Normal; all white except for black nose, ears, and feet, which makes it a white-pelted rabbit for practical purposes

MARKET: Number-two meat rabbit; often crossed with New Zealand White

Champagne D'Argent

IDEAL WEIGHT: 10–10.5 pounds (4.5–4.8 kg)

VARIETIES: One; the name means "French silver," and there are no other colors primarily for that reason

FUR TYPE: Normal

MARKET: A superb meat rabbit, with perhaps better dressout than New Zealands and Californians because of extremely fine bone. Fascinating color changes take place: young are born jet black; they soon begin to silver, with only the muzzle retaining the black upon maturity.

MEDIUMS

New Zealand

IDEAL WEIGHT: 10–11 pounds (4.5–5.0 kg)

VARIETIES: Black, Red, White

FUR TYPE: Normal

MARKET: Number-one meat and laboratory rabbit (White); converts feed to meat efficiently with fine bones; produces most consistently of all breeds. Widely available — a great first choice; a great only choice.

Palomino

IDEAL WEIGHT: 9 pounds (4.1 kg)

VARIETIES: Golden, Lynx

FUR TYPE: Normal

MARKET: Another excellent meat rabbit but not raised nearly as widely as New Zealands or Californians, partly because of the colors: many raisers find the colors attractive, but processors do not, preferring white. Very similar to New Zealand Red except for colors.

MEDIUMS

Rex

IDEAL WEIGHT: 8–9 pounds (3.6–4.1 kg)

VARIETIES: Amber, Black, Black Otter, Blue, Californian, Castor,
Chinchilla, Chocolate, Lilac, Lynx, Opal, Red, Sable, Seal, White,
Broken Group

FUR TYPE: Rex, or short and plushlike; similar to velour or velvet.
Beautiful fur on an excellent meat rabbit, with just about any color you
like, because almost all normal-furred breeds have been "rexed."

MARKET: Meat, primarily; fur, secondarily, if raised to maturity when pelt
is prime

PROBLEM: Poorly furred footpads can lead to sore hocks in hutches;
look for well-furred footpads

Satin

IDEAL WEIGHT: 9.5–10 pounds (4.3–4.5 kg)

VARIETIES: Black, Blue, Californian, Chinchilla, Chocolate, Copper, Opal, Otter, Red, Siamese, White, Broken Group

FUR TYPE: Satin; has hollow, luminous hair shaft that gives the breed a great deal of sheen

MARKET: An excellent meat rabbit, with pelt potential. A great number of normal-fur rabbits have been satinized, so you can obtain Satins in just about any color you like. An extremely popular rabbit at shows.

SMALLS

Dutch

IDEAL WEIGHT: 4.5 pounds (2 kg)

VARIETIES: Black, Blue, Chocolate, Gray, Steel, Tortoise

FUR TYPE: Normal

MARKET: Basically a fancier's rabbit, bred for exhibition, but an excellent small meat rabbit that dresses out well. Does are wonderful foster mothers, and some people keep a few just for that reason. It's a "panda bear rabbit," sometimes called Dutch belted, and its markings set it apart from every other breed. Markings are largely a matter of luck, so every new litter is eagerly anticipated. You never know for sure what you will get: mostly you get frustration and disappointment, but hope springs eternal, and once in a while you get a magnificent surprise — a beautifully marked Dutch.

English Angora

IDEAL WEIGHT: 5.5–6.5 pounds (2.5–3.0 kg)

VARIETIES: Agouti, Pointed White, Self, Shaded, Solid, Ticked

FUR TYPE: Angora, which is wool; primarily for spinning into yarn for luxurious garments

MARKET: Because it is smaller than other breeds of Angoras, it is less popular as a commercial animal, but its tufted ears give it a distinctive look admired by many.

SMALLS

English Spot

IDEAL WEIGHT: 6–7 pounds (2.7–3.2 kg)

VARIETIES: Black, Blue, Chocolate, Gold, Gray, Lilac, Tortoise

FUR TYPE: Normal

MARKET: Another fancier's breed. As a meat rabbit it is poor, having a racy, arched shape. It is a big challenge to the fancier, much as the Dutch is. The trail of spots along its sides must conform to a certain pattern that is not easy to achieve. You can drive yourself crazy with this breed — but devotees would raise no other.

Florida White

IDEAL WEIGHT: 5 pounds (2.3 kg)

VARIETIES: One

FUR TYPE: Normal

MARKET: A meat rabbit of small size that dresses out better than any other breed. It is basically a solid block of meat, with small ears, small feet, and small bones. At twelve weeks you can get as much dressed, edible meat from a Florida White as you can from a breed that's twice as big at eight weeks. Has large litters for a small breed; often used in crosses with larger rabbits to produce meat.

SMALLS

Harlequin

IDEAL WEIGHT: 7.5 pounds (3.4 kg)

GROUPS: Japanese, Magpie

VARIETIES: Black, Blue, Chocolate, Lilac

FUR TYPE: Normal

MARKET: An obscure breed for fanciers. It looks like somebody threw a few cans of paint at it. You have to see it to believe it and to decide whether you want to breed it to the point where it looks like you threw the right colors of paint in the right places.

Havana

IDEAL WEIGHT: 5–5.5 pounds (2.3–2.5 kg)

VARIETIES: Black, Blue, Broken, Chocolate

FUR TYPE: Normal but with sheen superior to most other normal-furred rabbits

MARKET: A somewhat obscure breed for fanciers but has devoted adherents; a pretty good small meat rabbit for home use

SMALLS

Mini Lop

IDEAL WEIGHT: 5.5–6 pounds (2.5–2.7 kg)

VARIETIES: Agouti, Broken, Pointed White, Self, Shaded, Solid, Ticked

FUR TYPE: Normal

MARKET: Sought as a pet by those who think lop ears are attractive; comes in many colors and patterns. Dresses out well; popular at shows.

Tan

IDEAL WEIGHT: 4–6 pounds (1.8–2.7 kg)

VARIETIES: Black, Blue, Chocolate, Lilac

FUR TYPE: Normal but with great sheen, similar to Havana, which doubtless is an ancestor

MARKET: Only for the fancier who demands a great challenge on the show table because of the many requirements of its markings and colors. The difference in breeding for a Tan's markings compared to Dutch or English, for example, is that you can breed for improvement and expect to get it.

DWARFS

American Fuzzy Lop

IDEAL WEIGHT: 3.5 pounds (1.6 kg)

VARIETIES: Agouti, Broken, Pointed White, Self, Shaded, Solid

FUR TYPE: Angora wool but shorter length than the larger Angora breeds

MARKET: Sought-after pet because of small size, lop ears, and wool in lots of colors

Britannia Petite

IDEAL WEIGHT: 2.4 pounds (1.1 kg)

VARIETIES: Agouti, Black, Black Otter, Chestnut Agouti, Ruby Eyed White, White

FUR TYPE: Normal

MARKET: Fanciers who want to show rabbits but don't have much space; a dainty little rabbit that likes to pose

DWARFS

Dwarf Hotot

IDEAL WEIGHT: 2.25 pounds (1 kg)

VARIETIES: White only, with big brown eyes and black eye band that resembles a heavy layer of mascara

FUR TYPE: Normal

MARKET: With its round fur-ball body, short ears, and come-hither eyes, this breed is a coveted pet.

Holland Lop

IDEAL WEIGHT: 3 pounds (1.4 kg)

VARIETIES: Agouti, Broken, Pointed White, Self, Shaded, Solid, Ticked

FUR TYPE: Normal

MARKET: As pets and for fanciers; small size, pug nose, many colors, and lop ears rank it high in the minds of pet rabbit owners.

DWARFS

Jersey Wooly

IDEAL WEIGHT: 3 pounds (1.4 kg)

VARIETIES: Agouti, Pointed White, Self, Shaded, Tan Pattern

FUR TYPE: Angora; longer than that on American Fuzzy Lop

MARKET: Wool, but doesn't have as much as larger Angoras; popular
pets because of small size, wooly appearance. Looks bigger than it is
because of all that wool.

Mini Rex

IDEAL WEIGHT: 4 pounds (1.8 kg)

VARIETIES: Black, Blue Eyed White, Blue, Castor, Chinchilla, Chocolate, Himalayan (Californian), Lilac, Lynx, Opal, Otter, Red, Sable Point, Seal, Tortoise, White, Broken Group

FUR TYPE: Rex

MARKET: As pets and for fanciers; an extremely popular little rabbit because of its marvelous coat in many colors

DWARFS

Netherland Dwarf

IDEAL WEIGHT: 2 pounds (0.9 kg)

VARIETIES: Officially Agouti, Self, Shaded, Tan Pattern but also Any Other Variety, which means just about any color or pattern you can imagine

FUR TYPE: Normal

MARKET: As pets and for fanciers. Very popular as pet and show rabbit, with apple-round head and short ears. A big surprise in the nest box with just about every litter — you never know what colors you will get.

Polish

IDEAL WEIGHT: 3.5 pounds (1.6 kg)

VARIETIES: Black, Blue, Chocolate, Blue-Eyed White, Ruby-Eyed White, Broken Group

FUR TYPE: Normal

MARKET: Fanciers primarily, as other dwarf rabbits seem more popular as pets. Has a great number of fans in the showroom and has been around longer than the other dwarf breeds.

GLOSSARY

abattoir. A slaughtering facility.

Agouti. Rabbit fur that has a salt and pepper appearance.

Angora. Rabbit with coat about 3 inches (7.5 cm) long. Raised for wool as well as meat.

breed. Race of rabbits distinguished by characteristics, such as color, size, body type.

breeder. An adult rabbit used for propagation. Also, a person who breeds rabbits in an effort to maintain and improve their special qualities. Breeders consider themselves more than just "raisers."

broken. Color pattern in which blotches of color appear on a white background.

buck. A male rabbit.

bunny. A cutesy term for rabbit. Babies are called bunnies for lack of another term, although some call the babies kits, even pups.

cobby. Short, stocky body type.

coccidiosis. A debilitating condition caused by an overload of a parasite living in all rabbits.

compactness. Refers to short, stocky body type or conformation.

condition. General state of physical well-being, revealed by such factors as brightness of eye, sheen of coat, and firmness of flesh.

crossbreeding. Mating animals of two different breeds.

cull. To eliminate a rabbit from the herd; a rabbit you don't want.

density. Thickness of coat of fur.

dewlap. Fold of loose skin under the chin of does; normal in some breeds but a disqualification in others. Not a disease condition.

disqualification. A temporary or permanent physical defect in a show rabbit. Listed in standard for each breed.

doe. Female rabbit.

dressout. Refers to percentage of edible meat available upon butchering.

dwarf. Rabbit weighing no more than 3 pounds (1.4 kg) at maturity.

ear canker. Scabby condition inside ear of rabbit; caused by mites.

fly back. Fur that returns quickly to normal position when stroked "against the grain." Desirable fur condition in many normal-fur breeds.

gestation. Period of 28–34 days (usually 31) from mating to birth of litter.

giant. Rabbit weighing 12 to 16 pounds (5.4 to 7.3 kg) or more at maturity.

guard hair. Coarser, longer hair than that of underfur.

hock. First joint of hind leg thickly padded with fur. What Thumper thumped.

hopper feeder. Metal feeder fastened to outside of the hutch with a trough that protrudes inside. Sometimes called a J-feeder or a self-feeder.

hutch. Rabbit cage; the best are constructed only of wire and metal.

inbreeding. Mating close relatives. Excellent practice when done wisely.

J clips. Special J-shaped metal clips used in hutch construction. Special pliers required for application.

junior. Rabbit less than 6 months of age.

kindle. Give birth to a litter.

linebreeding. Inbreeding for successive generations, generally mating "on a line" from and back to older specimens of the same family to capitalize on special characteristics and to breed them into descendants.

marked. Having a fur pattern of two or more colors, such as the Tan or the Dutch.

medium breed. Rabbit group with mature weight of 9 to 12 pounds (4 to 5.4 kg).

molting. Shedding fur.

mucoid enteritis. Severe diarrhea.

nest box. A place for the doe to give birth; provided on the 27th day after mating.

nick. A mating that produces a high percentage of good youngsters.

nonspecific enteritis. Diarrhea.

outcrossing. Breeding two unrelated animals of the same breed.

pair. Male and female to be mated to each other.

palpate. To feel for young in uterus of doe through abdomen. In this context, a pregnancy test.

pedigree. Record of ancestry. Should include at least three generations.

purebred. A specimen of a recognized breed from ancestors of that breed.

racy. Slim, slender body type. A Belgian Hare is racy; a New Zealand is not.

registration. Official examination of the rabbit and recording of rabbit pedigree by ARBA registrar. Only ARBA members may have rabbits registered.

Rex. Rabbits with short, plush fur.

runt. A baby rabbit markedly smaller than its littermates.

Satin. Rabbits with transparent hair shaft providing extremely lustrous coat.

scours. Diarrhea.

self. A rabbit that has the same color over its entire body.

senior. Rabbit over 6 months if of a breed that matures to a weight of less than 10 pounds (4.5 kg). Over 8 months if mature weight exceeds 10 pounds.

snuffles. Viral respiratory disease, highly contagious, marked by nasal discharge.

sore hocks. Ulcerated footpads.

specialty club. Rabbit club that specializes in a single breed.

standard. Written physical description of the perfect specimen of a breed.

strain. A family of rabbits within a breed and variety, exhibiting distinguishing characteristics and passing them on from generation to generation; sometimes called a line.

tattoo. To mark ears with a permanent identification mark or number. Private rabbitry number goes in left ear; registration number goes in right.

test mating. Returning doe to buck a week or 10 days after mating. If she complains, she may be pregnant. Palpation is a surer method.

thoroughbred. Purebred.

trio. A buck and two does of the same breed and perhaps the same variety.

type. Body conformation.

variety. A group within a breed and identified by color; for example, New Zealand Red is a variety of the New Zealand breed.

weaning. Separating the young from the doe, usually at 6 to 8 weeks of age.

INDEX

Page references in **bold** indicate charts; page references in *italics* indicate illustrations.

STOREY'S GUIDE TO RAISING SERIES

For decades, animal lovers around the world have been turning to Storey's classic guides for the best instruction on everything from hatching chickens, tending sheep, and caring for horses to starting and maintaining a full-fledged livestock business. Now we're pleased to offer revised editions of the Storey's Guide to Raising series — plus two much-requested new books.

Whether you have been raising animals for a few months or a few decades, each book in the series offers clear, in-depth information on new breeds, latest production methods, and updated health care advice. Each book has been completely updated for the twenty-first century and contains all the information you will need to raise healthy, content, productive animals.

Storey's Guide to Raising BEEF CATTLE (3rd edition)

Storey's Guide to Raising RABBITS (4th edition)

Storey's Guide to Raising SHEEP (4th edition)

Storey's Guide to Raising HORSES (2nd edition)

Storey's Guide to Training HORSES (2nd edition)

Storey's Guide to Raising PIGS (3rd edition)

Storey's Guide to Raising CHICKENS (3rd edition)

Storey's Guide to Raising MINIATURE LIVESTOCK (1st edition)

Storey's Guide to Keeping HONEY BEES (1st edition)

Storey's Guide to Raising DAIRY GOATS (4th edition)

Storey's Guide to Raising MEAT GOATS (2nd edition)

Storey's Guide to Raising DUCKS (2nd edition)

Storey's Guide to Raising POULTRY (4th edition)

Storey's Guide to Raising TURKEYS (3rd edition)

Storey's Guide to Raising LLAMAS

Other Storey Titles You Will Enjoy

The Backyard Homestead Guide to Raising Farm Animals,
edited by Gail Damerow.
Expert advice on raising healthy, happy, productive farm animals.
360 pages. Paper. ISBN 978-1-60342-969-6.

Farm Anatomy, by Julia Rothman.
A charming and richly entertaining visual guide to the curious parts and pieces of
rural living, dissecting everything from tractors and pigs to hay bales, farm tools,
and squash varieties.
224 pages. Paper with flaps. ISBN 978-1-60342-981-8.

The Fleece & Fiber Sourcebook, by Deborah Robson & Carol Ekarius.
A one-of-a-kind photographic encyclopedia featuring more than 200 animals and
the fibers they produce.
448 pages. Hardcover with jacket. ISBN 978-1-60342-711-1.

How to Build Animal Housing, by Carol Ekarius.
An all-inclusive guide to building shelters that meet animals' individual needs:
barns, windbreaks, and shade structures, plus watering systems, feeders, chutes,
stanchions, and more.
272 pages. Paper. ISBN 978-1-58017-527-2.

Rabbit Housing, by Bob Bennett.
Designs for building efficient facilities that can shelter two to one hundred
backyard rabbits in safety and comfort.
144 pages. Paper. ISBN 978-60342-966-5.

Small-Scale Livestock Farming, by Carol Ekarius.
A natural, organic approach to livestock management to produce healthier
animals, reduce feed and health care costs, and maximize profit.
224 pages. Paper. ISBN 978-1-58017-162-5.

The Spinner's Book of Yarn Designs, by Sarah Anderson.
Step-by-step instructions and photographs illustrate the process of turning fleece
into 80 distinctive yarns.
256 pages. Hardcover with 32 technique cards in an envelope.
ISBN 978-1-60342-738-8.

Successful Small-Scale Farming, by Karl Schwenke.
An inspiring handbook to introduce the small-farm owner to the real potential
and difficult realities of living off the land.
144 pages. Paper. ISBN 978-0-88266-642-6.

These and other books from Storey Publishing are available
wherever quality books are sold or by calling 1-800-441-5700.
Visit us at *www.storey.com* or sign up for our newsletter
at *www.storey.com/signup.*